Hermeneutics, History, and Technology

For better and worse, the future is often conceived in technological terms. Technology is supposed to meet the challenge of climate change or resource depletion. And when one asks about the world in 20 or 100 years, answers typically revolve around AI, genome editing, or geoengineering.

There is great demand to speculate about the future of work, the future of mobility, Industry 4.0, and Humanity 2.0. The humanities and social sciences, science studies, and technology assessment respond to this demand but need to seek out a responsible way of taking the future into account. This collection of papers, interviews, debates grew out of disagreements about technological futures, speculative ethics, plausible scenarios, anticipatory governance, and proactionary and precautionary approaches. It proposes Hermeneutic Technology Assessment as a way of understanding ourselves through our ways of envisioning the future. At the same time, a hermeneutic understanding of technological projects and prototypes allows for normative assessments of their promises.

Is the future an object of design? This question can bring together and divide policy makers, STS scholars, social theorists, and philosophers of history, and it will interest also the scientists and engineers who labor under the demand to deliver that future.

Armin Grunwald is Professor of Philosophy and Ethics of Technology and Director of the Institute for Technology Assessment and Systems Analysis at Karlsruhe Institute of Technology, Germany. His research fields include technology assessment and ethics of new technologies. He is author of multiple publications, including *Technology Assessment in Practice and Theory* (2018).

Alfred Nordmann is a Philosopher of Science and of Technoscience at the Technical University of Darmstadt. His current interests concern working knowledge and principles of composition as epistemological and aesthetic foundations of technoscience. He published introductions to Wittgenstein's *Tractatus and the Philosophy of Technology* as well as *Methodological Critiques of Technological Futures.*

Martin Sand is Assistant Professor of Philosophy of Technology at Delft University of Technology, the Netherlands. He investigates digital utopias utilizing recent insights from political philosophy. He teaches various courses on engineering ethics and is co-editor of the book series, *Futures of Technology, Science and Society*.

History and Philosophy of Technoscience
Series Editor: Alfred Nordmann

The Reform of the International System of Units (SI)
Edited by Nadine de Courtenay, Olivier Darrigol, and Oliver Schlaudt

The Past, Present, and Future of Integrated History of Philosophy of Science
Edited by Emily Herring, Konstantin S. Kiprijanov, Kevin Jones, and Laura M. Sellers

Nanotechnology and Its Governance
Arie Rip

Perspectives on Classification in Synthetic Sciences
Unnatural Kinds
Edited by Julia Bursten

Reflections on the Practice of Physics
James Clerk Maxwell's Methodological Odyssey in Electromagnetism
Giora Hon and Bernard Goldstein

Philosophy of Interdisciplinarity
Studies in Science, Society and Sustainability
Jan Cornelius Schmidt

Material Hermeneutics
Reversing the Linguistic Turn
Don Ihde

Post-Truth Imaginations
New Starting Points for Critique of Politics and Technoscience
Edited by Kjetil Rommetveit

Living in a Nuclear World
From Fukushima to Hiroshima
Edited by Bernadette Bensaude-Vincent, Soraya Boudia, and Kyoko Sato

Hermeneutics, History, and Technology
The Call of the Future
Edited by Armin Grunwald, Alfred Nordmann, and Martin Sand

Hermeneutics, History, and Technology

The Call of the Future

Edited by Armin Grunwald, Alfred Nordmann, and Martin Sand

Routledge
Taylor & Francis Group
LONDON AND NEW YORK

First published 2023
by Routledge
4 Park Square, Milton Park, Abingdon, Oxon OX14 4RN

and by Routledge
605 Third Avenue, New York, NY 10158

Routledge is an imprint of the Taylor & Francis Group, an informa business

British Library Cataloguing-in-Publication Data
A catalogue record for this book is available from the British Library

Library of Congress Cataloging-in-Publication Data
Names: Grunwald, Armin, editor. | Nordmann, Alfred, 1956- editor. | Sand, Martin, editor.
Title: Hermeneutics, history, and technology : the call of the future / edited by Armin Grunwald, Alfred Nordmann, Martin Sand.
Description: Abingdon, Oxon ; New York, NY : Routledge, 2023. | Series: History and philosophy of technoscience ; 24 | Includes bibliographical references and index.
Identifiers: LCCN 2022058395 (print) | LCCN 2022058396 (ebook) | ISBN 9781032344652 (hardback) | ISBN 9781032344669 (paperback) | ISBN 9781003322290 (ebook)
Subjects: LCSH: Technology assessment. | Hermeneutics.
Classification: LCC T174.5 .H465 2023 (print) | LCC T174.5 (ebook) | DDC 303.48/3--dc23/eng/20230208
LC record available at https://lccn.loc.gov/2022058395
LC ebook record available at https://lccn.loc.gov/2022058396

ISBN: 978-1-032-34465-2 (hbk)
ISBN: 978-1-032-34466-9 (pbk)
ISBN: 978-1-003-32229-0 (ebk)

DOI: 10.4324/9781003322290

Typeset in Times New Roman
by MPS Limited, Dehradun

Contents

Figures

Contributors

Vera Borrmann is a Research Associate in the project *Agency and Neurotechnology* of the *SALTUS! Group Responsible Artificial Intelligence* at Freiburg Institute for Advanced Studies (FRIAS) at the Albert-Ludwigs-Freiburg, Germany.

Christopher Coenen is Research Group Leader of the research group *Health and Mechanization of Life* at the Institute for Technology Assessment and System Analysis at Karlsruhe Institute of Technology, Germany. He is coordinator of the transnational EU-ERA-NET-Project *FUTUREBODY* and leading editor for the journal *NanoEthics*.

Sascha Dickel is Professor for Sociology of Media and Theory of Society at Johannes Gutenberg University Mainz, Germany. He is co-speaker of the DFG research group "Practicing Evidence – Evidencing Practice" and was head of the BMBF funded collaborative project "Prototyping Futures."

Steve Fuller holds the Auguste Comte Chair in Social Epistemology in the Department of Sociology at the University of Warwick, England, where he is the founder of the research program of social epistemology. He is also a Senior Research Fellow at the Breakthrough Institute, and Affiliate Scholar at the Institute for Ethics and Emerging Technologies.

Christopher Groves is Research Fellow at the School of Social Sciences at Cardiff University, UK. Since 2016, he has worked on the Flexible Integrated Energy Systems (FLEXIS) Project. He is also author of the book *Care, Uncertainty and Intergenerational Ethics* and a new book is to be published in 2023, provisionally entitled *How We Suffer: Towards an Understanding of Collective Injury*.

Armin Grunwald is a Full Professor of Philosophy and Ethics of Technology at Karlsruhe Institute of Technology (KIT), Germany. He is Director of the Institute for Technology Assessment and Systems Analysis at KIT and Director of the Office of Technology Assessment at the German Bundestag.

Sheila Jasanoff is Pforzheimer Professor of Science and Technology Studies at the Harvard Kennedy School, USA. She is author or editor of more than 15 books, including *The Fifth Branch, Science at the Bar, Designs on Nature, The Ethics of Invention,* and *Can Science Make Sense of Life?*. She also founded and directs the STS Program at Harvard University.

Arthur Wei-Kang Liu studies physics, philosophy, and political science at the Technical University of Darmstadt, Germany, with special interest in the philosophy of science. As a Student Research Assistant he participated in the discussions of hermeneutic technology assessment.

Alfred Nordmann was Professor for the Philosophy of Science and of Technoscience at the University of South Carolina, USA, and Technical University of Darmstadt, Germany. He is a guest professor at Peter the Great Saint Petersburg Polytechnic University, Russia, editor of the book series *History and Philosophy of Technoscience* and of the journal *Technology and Language*.

Martin Sand is assistant professor of ethics and philosophy of technology at Delft University of Technology, the Netherlands. He teaches courses in engineering ethics and researches technological futures and utopias.

René von Schomberg is a Senior Research Fellow at Käte Hamburger Kolleg *Cultures of Research* at the RWTH Aachen University, Germany, Guest Professor at the Technical University of Darmstadt, Germany, and Former Official of the European Commission. He is author and (co-)editor of 14 books, including the *International Handbook on Responsible Innovation, a Global Resource*.

Preface

> Are we speaking about something in the future or something in the present when we point at the sky and say "It will rain soon"?
> (Ludwig Wittgenstein, *Philosophical Occasions*, 417)

The future is in great demand. But are we speaking about the future or the present when we say "Due to climate change, ocean-levels will be rising soon" and are we speaking about the present or the future when we say "People will soon control machines by thought alone"? Wittgenstein's remark is challenging and sobering at once, challenging us to consider how the future appears in our talking and thinking about technology.

The future is in great demand. Fairs and exhibitions, popular and academic publications are dedicated to Industry 4.0 and Humanity 2.0, to the Future of Mobility or the Future of Work. In the absence of ideological struggles and utopian visions for a pursuit of the good life, these presentations of the future foreground technology. Indeed, there is a tendency to reduce the future to technological development in that public discourse focuses on the future as somehow emanating from the laboratories of science and industry – be it AI robotics or the hydrogen economy, be it vaccines and magic bullets in the treatment of disease, weapons and monitoring systems for conflict resolution. Inversely, technology is often seen as the answer to the central questions of the future: Politicians, Climate Panel reports, and TV documentaries consider technological innovation as a last hope for saving our planet from climate change, environmental pollution, or resource depletion. The future is calling! For better and worse, the future is increasingly defined as a set of problems requiring engineering or management solutions – e.g. in the guise of global challenges, Millennium and Sustainable Development Goals (MDGs and SDGs), or the need for CO_2 reduction. Science and technology are now called upon to meet these challenges and to prevent a catastrophic future. And simultaneously, for better or worse, we discuss, how AI or the hydrogen economy will transform our world. In the first case, a future development is taken as a given and one awaits how technology might respond to it. In the second case, a technological development is taken as a given, raising the question how it might affect societies and our planet in the future.

In the midst of these discussions arise several methodological quandaries and confusions. This volume is dedicated to these and proposes a hermeneutic approach to deal with them. One of these confusions arises with the inability or unwillingness to distinguish two perfectly sensible and one almost unintelligible question. Perfectly sensible is the question *Where are we heading?* – it concerns the extrapolated, perhaps catastrophic future of our socio-technical world if present trends continue. Perfectly sensible but quite distinct is the question *How shall we live?* – it concerns in aspirational terms what kind of world we should be striving for. Almost unintelligible, in contrast, is the question *How will it be once the future has changed us?* – this question seeks to imagine something that, by definition, lies beyond our reach, namely a whole new world that arose along with novel technological capabilities. It is crucial to understand how and why these questions have become conflated, and to disentangle them.

Such quandaries and methodological challenges have formed the backdrop of an academic conversation that has been reconstructed in the opening chapter of the present volume (Arthur Wei-Kang Liu). Referring to Enlightenment conceptions of history and reflections about the limits of knowledge and power, one of the editors (Alfred Nordmann), critiqued in a series of papers what he called "speculative ethics." This critique was extended to a technological hubris in the Collingridge Dilemma, certain scenario-building methodologies, or "anticipatory" Technology Assessment (TA). Representing a highly reflective institutional practice, the TA researcher and co-editor of the present volume, Armin Grunwald, argued for the necessity of understanding the desirability of technological futures in order to orient present-day decision-making. Over the course of time, the discussion of these complementary positions foregrounded that, at any rate, "the future" is not a referent but a figure of discourse which amalgamates present-day hopes and desires, fears and anxieties, aspirations and anticipations. From this conversation arose the conception of "hermeneutic TA."

At this place of convergence, however, new questions and disagreements emerged on how to respond to the call of the future and how to formulate a program for hermeneutic TA. Is this a project merely of self-understanding through discourse analysis or does it have normative dimensions which would allow us to actually assess emerging technologies? Is it addressed only to linguistic expressions or can there be a hermeneutics also of technological systems, artefacts, models, and technical practice? Is it a radical departure from the misplaced ambition and technological hubris of taking the future as an object of design or control, or is it a valuable addition to the familiar toolbox of TA, providing a humanities perspective to complement technical and social science dimensions? Is the knowledge gained from studying "the future" as an amalgamation of present-day hopes and fears policy-relevant and if so, how?

These questions are by no means resolved even among the three editors. At the same time, the urgency and cultural significance of these questions was highlighted by the vigorous engagement with them in a series of workshops involving prominent thinkers about science and society: Sheila Jasanoff, René von Schomberg, Steve Fuller, Arie Rip, along with George Khushf and Christopher Groves, Christina Schües and Sebastian Pfotenhauer, Ben Hurlbut and Sascha Dickel, Christopher Coenen and Vera Borrmann. The proposed volume is no mere anthology but designed to draw the reader into an ongoing conversation. It therefore assembles a programmatic statement of what hermeneutic TA could be and why it is needed, several papers with extended case-studies, an interview with Sheila Jasanoff, a synoptic reconstruction of topical conversations and debates, a debate on precautionary and proactionary attitudes. These chapters stand for themselves but are also tightly cross-referenced, allowing readers to make discoveries and extend the discussion.

As we invite our readers to reflect the future as a discursive figure or promising world, we want to thank Arthur Wei-Kang Liu for his contribution to the editorial process. Thanks are due also to the Thyssen Foundation for supporting one of our two workshops in Karlsruhe.

Part I

Debating the Program

1 On the Road to Hermeneutic Technology Assessment – A Historic-Systematic Reconstruction

Arthur Wei-Kang Liu

Introduction

The future is something we are all concerned with, whether it is something mundane like our next vacation or concerns the salient questions about human enhancement (Humanity 2.0), nanotechnology, the future of work, etc. All of them contain a technological dimension (see e.g. Dickel and Schrape 2015) and especially the latter ones gain ever-increasing political importance. It is as if the future is inseparably connected to the development of technology. One expression of this link is the Collingridge dilemma, which offers a riddle and poses a challenge for the mastery of "emerging technologies" with regard to their promise for the future. Furthermore, these technological promises and anticipations carry a whiff of utopian thought, they entail visions of betterment for all sorts of social, ecological, and political problems. This notion of *the* future was soon challenged by Science and Technology Studies (STS) thinkers, Technology Assessment (TA), Futures Studies, and philosophers (of technology). The present introduction sheds some light on our thinking about and handling of the future by reconstructing this discussion, which eventually gave rise to the demand for a hermeneutic TA. I will comment on key notions and passages from the original contributions to this debate and identify the main arguments presented there. This will result in a collage of a dynamic and ongoing discussion that spans across various fields and disciplines, to which the present book also testifies and which it furthers. My reconstruction sets a focus on the dialogue in the course of nanoethics and human enhancement between Alfred Nordmann and Armin Grunwald, while considering their origins and also valuable impulses that have contested, advanced, and transformed their views by Sheila Jasanoff, Rebecca Roache, Cynthia Selin, and others. At the end of this chapter, I outline the possibilities of a hermeneutic TA that have presented themselves in this discussion, anticipating some key aspects of the debate that unfolds in the following chapters, and I will also provide a first outlook for possible future developments of hermeneutic TA. In general, this introduction together with the contribution to the present

DOI: 10.4324/9781003322290-2

volume shall serve as an invitation to pick up on these strands and ideas and further develop and refine the idea of a hermeneutic TA.

Technologies of Hubris

The starting point of our discussion is Sheila Jasanoff's plea for "technologies of humility" (Jasanoff 2003, 227). In her essay, she analyzes the relations between decision-makers, experts, and citizens in the management of technology. Among other things, she underlines the essential role of "predictive methods" for the interaction between the three mentioned groups. She states that there is

> a series of predictive methods (e.g. risk assessment, cost-benefit analysis, climate modelling) that are designed, on the whole, to facilitate management and control, even in areas of high uncertainty. These methods achieve their power through claims of objectivity and a disciplined approach to analysis, but they suffer from three significant limitations. First, they show a kind of peripheral blindness toward uncertainty and ambiguity. Predictive methods focus on the known at the expense of the unknown, producing overconfidence in the accuracy and completeness of the pictures they produce. [...] Second, the technologies of predictive analysis tend to pre-empt political discussion. Expert analytic frameworks create high entry barriers against legitimate positions that cannot express themselves in terms of the dominant discourse. Claims of objectivity hide the exercise of judgment, so that normative presuppositions are not subjected to general debate. Third, predictive technologies are limited in their capacity to internalize challenges that arise outside their framing assumptions. (Jasanoff 2003, 238–239)

This observation brings us to the heart of the criticism about predictive methods and, respectively, TA. As the name implies, predictive methods aim for predictions, for example, about future trends and consequences of technologies or decisions concerning them, even though those predictions are sometimes linked to high uncertainty with regard to their ostensible precision, accuracy, completeness, and objectivity. For this reason, she calls them "technologies of hubris" (Jasanoff 2003, 238); they promise more than what they can possibly deliver. Despite the insufficiency of the method, the public, experts, and, in particular, politicians use them in good faith to justify policy decisions or policy recommendations as if it is a simple matter of human choice and action to realize them. Therefore, Jasanoff calls for the use of "technology of humility" because they

> complement the predictive approaches: to make apparent the possibility of unforeseen consequences; to make explicit the normative that

> lurks within the technical; and to acknowledge from the start the need for plural viewpoints and collective learning. [...] [Framing, vulnerability, distribution and learning] provide a framework for the questions we should ask of almost every human enterprise that intends to alter society: what is the purpose; who will be hurt; who benefits; and how can we know? (Jasanoff 2003, 240)

Evidently, she underscores the urgent need for normative reflexion on technologies and their implicated societal changes. In general, she challenges the notion of TA along with the expectations for it as a way to anticipate *the* future.[1] She demands a humbler and more sceptic attitude toward the possibility of predicting the future as well as an acknowledgement of our epistemological limitations.

Ethics, the Future, and Nanotechnology

In the meantime, nanotechnology was discussed widely among the public. Works such as "Nanotechnology: Shaping the World Atom by Atom" (Amato 1999), published by the National Science and Technology Council of the USA, provoked many responses and critical concerns regarding the ambitions of nanotechnology like human enhancement, which was one of its most pronounced goals for improvement (Nordmann 2007a, 159).

The Case against Speculative Nanoethics

In 2007, Alfred Nordmann reacted to the numerously published ethical enquiries concerning nanotechnology and human enhancement – the so called "emerging" or "converging" technology – which are obviously future-oriented. Above all, he criticizes the high-speculative foundations of those analyses and discussions concerning these fields. He diagnoses a problematic "'if and then' syndrome":

> An if-and-then statement opens by suggesting a possible technological development and continues with a consequence that demands immediate attention. What looks like an improbable, merely possible future in the first half of the sentence, appears in the second half as something inevitable. And as the hypothetical gets displaced by a supposed actual, an imagined future overwhelms the present. (Nordmann 2007b, 32)

In other words, speculative ethics is concerned with the implications of a possible, but very improbable "if" which demands immediate action to prevent its realization. The probability of the conditional is not the subject of judgement anymore, because the consequence seems inevitable anyway. As he puts it, "considerations of the present are overwhelmed by the supposed imminence of a highly speculative future" (Nordmann 2007b, 39). By analyzing those arguments in detail, Nordmann identifies three types of "arguments" (33) used for justification:

- Foreshortening: "[T]he most suggestive 'argument' for a hypothetical future that is upon us already comes from a large family of logarithmic plots that extrapolate an accelerating speed of technical development from the past via the present into the future. If past and present trends continue [...], even seemingly remote technical capabilities will be upon us before we know it. These graphs not only extrapolate from the more or less recent past into the future but are themselves the result of an extrapolation from the computer industry's 'Moore's Law' to all of technology" (Nordmann 2007b, 35).
- Conflation: "They extrapolate from the past to the future by making all technologies look alike [in the sense of their purpose to overcome human limitations]" (Nordmann 2007b, 36–37).
- Straw man: "One either believes in optimal parameter values (that is, in a non-improvable human nature) or one should endorse cognitive enhancement" (Nordmann 2007b, 38–39).

Foreshortening, in this sense, is an argument created by technology for technology. It takes technology as a continuous development which just goes on and on. (This motion of thought can also be discovered in the expression of "emerging technologies," which will be elaborated later on [section "Recapitulation"].) Conflation, on the other hand, is an interpretation of technology which aims to construct a narrative that fits the current visions of technological development. To illustrate this, Nordmann takes Arthur Caplan's (2006, as quoted in Nordmann 2007b, 37) perspective on technology and human kind as an example. The argument states that "one either accepts the advent of technologically enhanced people of tomorrow or denies the obvious truth that humans have always used technology and thereby improved their condition" (Nordmann 2007b, 37). According to Nordmann, this argument fails to acknowledge two things: the subtle but categorical difference between managing and transcending as well as between transhumanism and individual enhancement as improvement. Moreover, these arguments fall short of recognizing the fact that technologies change society – e.g. its values – and vice versa; a projection of our values today on past developments and technological inventions is fallacious. Additionally, the notion of human enhancement is often denoted as a symbol of liberation from limitations (transcending limits), but actually it creates new (technological) dependencies. The straw man argument generates false dichotomies which force one – if one accepts them – to react to future visions of technology as if they were accurate predictions. Here "the burden of proof" (Nordmann 2007b, 39) is reversed. Normally, the person who claims must provide proof of his/her claim, but in the case of emerging technologies this burden seems reversed; you will be condemned for being reckless and ignorant of the "facts." Visions of the future which use the argumentative strategies above are nothing more than pure speculations.

An early paper of Armin Grunwald's (2007) on the issue of human enhancement provides another entry point: He considers human enhancement as a converging technology worth of ethical consideration, but acknowledges it as a "future projection" (3) instead of a factual prediction of the development of mankind because of the uncertainty involved in it. He also makes clear why we should, despite the partially fictional character of those visions, discuss and evaluate them. Even if they never realize or "falsely" present something as desirable,[2] they indeed affect opinions along with the ideas of the scientific community and the public. Thus, ethics and science should have a genuine interest in analyzing and dealing with futuristic visions:

> Occupation with futuristic visions, speculations and risks is not only sensible, but necessary. While their content often is near to fictions and might be rather far away from facts it is a fact that visions can have real impact on scientific and public discussions. We have to distinguish between the degree of facticity of the content of the visions and the fact that they are used in real communication processes with their own dynamics. (Grunwald 2007, 10; see also 2013)

His analysis of the problem of the perception of future visions overlaps with Nordmann's but underlines the conflation of normative and epistemic elements, so that both elements become in some way opaque:

> [I]n futuristic visions as in the case of human enhancement, there are normative elements (like the desire to overcome illness and death) and epistemic elements (like assumptions about what scientific knowledge will be available within a certain time frame) [...] intermixed. (Grunwald 2007, 9)

Nonetheless, he attributes the function of "orientation" to those ethical debates, since they "offer criteria and procedures to deal adequately and responsibly with the normative aspects and implications of these new occasions of choice" (Höffe 1993, as quoted in Grunwald 2007, 1) which emerged, but overall those future projections are, hence, just speculative:

> Future prospects on converging technology, therefore, show the maximally imaginable disorientation: they oscillate between expectations of paradise and of catastrophe. [...] [D]ue to the fact that, because of their speculative character, there are hardly any arguments for deciding "rationally" for one or the other variant, a serious problem

> arises here for the "orientational function" of communication of the future. (Grunwald 2007, 8)

It seems, therefore, even stranger that pure visions of nanotechnology are sometimes treated more seriously in ethical debates than current trends like climate change, which have higher scientific credibility. This motivates Nordmann to propose some differentiations and solutions. For him, there are two different framings for technology, which are connected to one question posed by each: what will technology do with us, or what can technology do for us? The first one is the notion Nordmann criticizes, due to its claims about a kind of technological determinism:

> However, if one embraces the challenge to evaluate what the future will bring, the very possibility of an ethical perspective on emerging technologies is undermined as one is left with a series of equally untenable positions. The first of these posits an eternally fixed human nature against which future realities might be measured. [...] Aside from its historical implausibility it can always be accused of making an unwarranted metaphysical assumption. Secondly, one could take the status quo as an arbitrary basis of judgement – but the way in which the question is framed informs us that this basis holds no longer and that the status quo is quite irrelevant for the future that is to be evaluated. Another option would therefore be to adopt a blatantly paternalistic attitude [...]. If one wants to avoid such paternalism at all costs, one could finally adopt a laissez faire attitude and refrain from ethical judgement. (Nordmann 2007b, 40)

Similarly Deborah G. Johnson (2007) argues in the same issue of the journal *NanoEthics* for a refusal of technological determinism from a STS perspective. Technological development is not an independent process; it unfolds in a social setting:

> Technology does not have a natural or logical order of development. Technological development is contingent; it is contingent upon a wide range of factors, internal and external to science and engineering. Scientists and engineers look at nature through a lens of human interests and human-made theories and concepts; engineers build things that fit into particular social and cultural contexts. In other words, scientists and engineers do not work in isolation; their work is shaped by social, cultural, economic, and political elements, as well as by what they learn about nature. (Johnson 2007, 24)[3]

After dismissing technological determinism, Johnson claims that nanoethics, or ethics of emerging technologies in general, can have a radically more active and leading role within the process of technology development. Once the notion of the future gets liberated from metaphysically laden premises, it is more likely to "shape" or "influence" the future in the sense of current developments. We can now ask, "[w]hy should we now accept this or that promise of a technological future?" (Nordmann 2007b, 40) and "nanoethicists can both disclose the values that are being constructed or contested and do the work of normative evaluation, asking what values should or should not be constructed?" (Johnson 2007, 28). It is not technology that can lead ethical reasoning. On the contrary, it is ethical reasoning and thought that can lead technological development. Nordmann concludes:

> Rather than adopt a believing attitude towards the future, an ethics beholden to present capabilities, needs, problems, and proposed solutions will begin with vision assessment [...]. Envisioned technologies are viewed as incursions on the present and will be judged as to their likelihood and merit: How credible are these claims, and do these technologies solve acknowledged problems? More generally: What do these visions tell us about the present, what is their implicit criticism of it, how and why do they require us to change? (Nordmann 2007b, 41)

Vision Assessment

Before digging further into the proposed handling of visions by Nordmann, the concept of vision assessment has to be explained in greater depth because this concept, as I demonstrate, dominated parts of the subsequent discussion. Grin and Grunwald (2000) proposed vision assessment as an additional tool of the repertoire for TA. It should fulfil three main tasks:

- Analysis: "With respect to analysis [...], it would be a question of disclosing the cognitive and normative contents of the visions and of judging epistemologically the extent of their reality and practicability, self-evidently on the basis of current knowledge. Then, an important aspect, the prerequisites for the visions' realizability and the time-spans involved, would have to be investigated" (Grunwald 2007, 11).
- Evaluation: "These are questions of how the cognitive aspects are to be categorized, how they can be judged according to the degree of realization or realizability, according to plausibility and evidence, and which status the normative aspects have, e.g., relative to established systems of values, or to ethical standards. The purpose is the transparent disclosure of the relationship between knowledge and values, between knowledge and the lack of it, and the evaluation of these relation" (Grunwald 2007, 11).

- Management: "[I]t is a matter of deciding and acting […]. The question is how the public, the media, politics, and science can be advised with regard to a "rational" use of visions. […] [A]n early ethical dealing with visions of the type of 'human enhancement' would be an important and highly relevant task in order to allow more transparent debate, especially about science's agenda" (Grunwald 2007, 12).

Vision assessment is, thus, meant to dissect and explicate the different components of a vision so that societal agents can decide rationally on issues connected to envisioned futures of a certain technology. A 'decision based on rationality' in this context refers to decision made within a discourse that is open and intellectually as well as conceptually accessible to everyone who wants to participate. This should ensure identification of the best means to fulfil certain aims, goals or values, on which all participants of the discourse can agree.[4] Vision assessment shall therefore contribute, above all, to a transparent and clarifying analysis of the cognitive and value-laden components of visions and their relation:

> Such an epistemology [vision assessment] would uncover and analyse the premises and presuppositions of the various types of inputs which constitute the resulting statement about future developments in order to allow for criticism of those inputs with respect to their validity as well as with respect to their composition. […] In particular, it would be the assignment of Vision Assessment in ethical regard to confront the various and, in part, completely divergent normative aspects of the visions of the future directly with one another. […] (Grunwald 2007, 9–10)

It shall, hence, enable a well-argued and reasoned decision in the end. As mentioned above, vision assessment obtains thus an "orientational function" by providing a critical dissection of all components involved. Now, coming back to Nordmann's proposed way of dealing with envisioned futures, we can see that he is convinced of the possibility of judging a vision's credibility and of the existence of implicit criticism of the present carried by the vision, by using vision assessment. Nordmann suggests three strategies to prevent foreshortening. The first one is that "the scientific community [is to be held] to its own standards of honesty and clarity" (Nordmann 2007b, 43). Secondly, he emphasizes the difference between physical and technological possibility.[5] Lastly, he highlights, in the context of nanotechnology and human enhancement, the categorical difference between engineering *for* mind or body and *of* mind or body. Following those pieces of advice, we are "reclaiming the present" (44) and the potential of technology has to be fully utilized by society.

To be clear, he does not criticize thought experiments as a tool widely used in philosophy (for example, the brain in a vat), because there is a fundamental difference between a thought experiment and the genuine belief that something is – or will be – the case:

> Indeed, if science fiction scenarios lead to interesting philosophical questions, it is precisely because one suspends disbelief in the presence of fiction. Relieved of the pressure to determine what is true or false, what is likely to happen and what not, we can forge ahead and explore who we are, who we might wish to be, and how these wishes reflect on ourselves or our views of human nature. (Nordmann 2007b, 43)

Consequently, he is not criticizing public debate about visions of technologies like human enhancement. Instead, his critique is "directed at those who discuss the pros and cons of human enhancement technologies *as if such technologies were upon us already*" (Nordmann 2007b, 42, italics in original). Exactly the lack of credibility of those speculative visions of the future and the dishonesty about it disqualify it for serious consideration for practical action; we cannot judge a technology ethically based on what it might achieve in the future, because we cannot know if it will really achieve it. Those debates can, therefore, only serve the purpose of reflecting on ourselves and the present. Because of the speculative nature of those discussions, Nordmann argues that speculative ethics attracts too much attention and "squanders" (Nordmann 2007b, 31) the already scarce resources dedicated to ethical evaluation as a result (argument of opportunity costs). Furthermore, these visions and the discussions surrounding them can distract us from more pressing present issues (argument of influence). Hence, we should stop to spend time on speculative scenarios in order to deal with more pressing present issues. In a later paper, co-authored with Arie Rip, these arguments were further sharpened:

- Argument of opportunity costs: "we can only do so much when our resources (ethical and otherwise) are limited, and other ethical questions may be even more important. There are good reasons to think that the opportunity costs of speculative ethics are too high, with less spectacular but more pressing 'here and now' ethical issues not getting the attention they deserve, and more speculative visions not being subjected to reality checks" (Nordmann and Rip 2009, 273).
- Argument of influence: "Speculative ethics poses a twofold danger: present developments are not questioned because no one is paying attention to them, and worries about the most futuristic visions of nanotechnology can cast a shadow on all ongoing work in nanoscience and technology" (Nordmann and Rip 2009, 274).

Rip and Nordmann are certainly aware that there are visions that will be realized in the future. Thus, an early evaluation could indeed pay off.

However, this is really improbable if one recalls that there are so many different, in some cases widely divergent, visions of the future. The opportunity costs are, therefore, really high because we will not be able to evaluate present issues, which will *certainly* pay off. The other argument, the argument of influence, is also linked to opportunity costs, insofar as speculative visions distract us from present issues. Furthermore, visions of technology have, as stated above, a real impact on agenda-setting, decision-making and the public opinion. In this sense, speculative visions could harm current and desirable developments of technology by overshadowing them, for example, through an improbable, yet not impossible apocalyptical vision. As a consequence, we should focus on the present and leave visions of the future, which are purely speculative, as a source for ethical consideration for policy recommendation behind us.

However, this does not mean that we cannot think about or "influence" the future. There are trends pointing toward the future, or perhaps foreshadowing the future that can be extrapolated from the present. We just need to keep in mind, following Nordmann, that we can only influence the trend but not *the* future itself:

> After all, there are also reasonable extrapolations of current developments and some slippery slope is perhaps more real than another. With its consistent orientation to past and present, to programs, presuppositions, and implications rather than envisioned consequences and future deliverances, this paper has tried to show that it is not important whether one can distinguish better and worse modes of scenario-construction: The "reasonable extrapolations" can be identified without looking at the extrapolated scenarios. One can recognize them by seeing, for example, that the scenarios underwrite currently funded research. (Nordmann 2007b, 44)

To summarize, Nordmann suggests that "[i]f supposedly toughminded foresight is premised upon credulity and intellectual sleight of hand, we ought not to take it seriously" (Nordmann 2007b, 34) for practical consequences or advice. Instead, we should judge each vision by its credibility as well as its probability on the basis of hard scientific evidence – vision assessment is an adequate instrument for this purpose. If the vision is based on evidence, we should take it seriously for ethical analysis, otherwise we consider what it says about us, about the present in which we live and learn for technological development.

Two Responses – Biased Credibility and Opportunity Costs

Of course, Nordmann's proposal and conclusions provoked many reactions of which two noteworthy ones shall be discussed. In a following volume of *NanoEthics*, Rebecca Roache (2008) responded directly to

Nordmann's paper, disagreeing with his analysis of speculative nanoethics. To begin with, she is convinced that "we cannot do away completely with considering speculative future possibilities, since doing so would leave us unable to address pressing current issues. Consider that any decision about which course of action to take will involve attending to possible future scenarios" (Roache 2008, 319). She gives us an example for her argument:

> My relatively trivial decision about whether to make myself a cup of tea in the next 5 min involves cogitations that, were they fully conscious, might be expressed along the lines of, 'If I don't make a cup of tea in the next 5 min, then I will probably grow parched and irritable', and, 'If I make a cup of tea in the next 5 min, then I may lose the thread of the paragraph I'm currently writing and will need to start again'. For all I know, the future scenarios to which I attend when deciding whether to make a cup of tea may be no more or less probable than those to which we attend when predicting the long-term prospects for human enhancement, and yet we all attend to such scenarios many times each day, without believing that, in general, doing so distracts us from more important activities. Further, since any decision about which course of action to take involves attending to possible future scenarios, and addressing current issues involves—among other things—making decisions about which course of action to take, attending to possible future scenarios can hardly be thought to take place at the expense of addressing current issues. (Roache 2008, 319–320)

It is worth noting that she wants us to consider "*possible* future scenarios" in order to be able to engage in present actions. However, she compares the *probability* of the *possible* scenarios surrounding human enhancement with the *probability* of the *possible* consequence of the "teacup decision." Indeed, one cannot deny the possibility of visions of human enhancement, but their probability (and maybe also possibility) cannot as easily be ascertained. Estimating the probability of the consequences of my "cup-of-tea-decision," on the contrary, is significantly easier because I experienced the constellation of consequences along with its circumstances before, which is why they came to my mind in the first place. This is exactly the point: Visions of human enhancement, for instance, take a leap in time to a future, which will be completely different from the present, the world as we know it. Since it is completely different from our world, we cannot really judge the possibilities and probabilities of the envisioned future and its effects. Interestingly, in what seems like a contradiction, Roache later claims that "[t]he most exciting future technologies are novel and often achieve what, previously, was barely dreamed possible" (Roache 2008, 322) – which actually supports the argument above about the impossibility of a serious evaluation of the possibility and probability of visions of the future.

In a direct response to Nordmann (2007b) as well as Nordmann and Rip (2009), Grunwald draws a similar conclusion: It is "impossible to decide *in advance* which concerns are justified and which are not. Decisions of this type are confronted with high uncertainties, and as history shows, full of surprises and unexpected developments" (Grunwald 2010, 95, italics in original). Roache phrases this argument, I call it the "argument of biased credibility," as follows:

> [T]he claim that ethical concern is squandered when applied to scenarios that are not 'credible' ([Nordmann 2007b, 43; Weckert and Moor 2007, 143]) is highly controversial, and may well be misconceived. That humans are often not very good at judging the probability of a given scenario is well-documented [...], meaning that to dismiss certain possible futures due to their incredibility is a risky strategy: some of them could turn out to be much more likely than we currently think. (Roache 2008, 323)

This argument brings us to the main point or the limitation of TA and, thus, also of vision assessment. TA and vision assessment, respectively, can just tell us what could happen bound to the available knowledge at the point in time of assessment. We are beholden to the knowledge that we have – nothing more nor less. Possibilities and probabilities can be re-evaluated any time as/when something unexpected or new occurs. We are reminded of the argument above, and indeed it is the argument of a biased perceptions of possibilities and probabilities in favor for the status quo in another phraseology. Consequently, we have to dissociate ourselves from the idea that we can easily judge a vision's credibility. In other words, "[t]he valid criticism points to a problem of a possible 'pure speculativeness' of speculative nanoethics, in which statements might run the danger of being *purely arbitrary*" (Grunwald 2010, 95, italics in original).

Taking the biased credibility as a starting point, Roache presents another argument. According to her, visions could entail motivational function: "the attempt to render a solution technically possible is motivated by the ethical judgment that such a solution would be a great good" (Roache 2008, 323). She is convinced that Nordmann's view does not allow allocating resources to an idea such as curing cancer, which was in the beginning just a vision and not supported by scientific evidence.

Moreover, Roache presents two arguments for speculation in the case of ethics of emerging technology which are connected to the argument of opportunity costs in general. The first one concerns the "ris[k] [of] squandering scientific effort" (2008, 322). She agrees that "scientific expertise is [a] scarce resource," so if ethical evaluation is constricted to only technological projects that seem to be possible to realize, two problems could arise. First, engineers and scientists could develop a technological solution for an ethical problem, which turns out not to be an ethical problem at all or that the technology does not solve the problem. Either of those outcomes would have squandered scientific effort. One can argue, nonetheless, that Nordmann did not suggest that ethics should not concern itself with visions of the future at all, as Grunwald recognizes too (2010, 93–94); ethicists should evaluate visions and their premises in order to influence current developments and to judge which problems *are* problems and how present technology could tackle them.[6] The other problem arises when ethicists evaluate technologies only if they are (almost) certainly realizable. Then it could be too late to stop their realization, because developers have a genuine interest to realize these technologies in which they invested time and other resources (Roache 2008, 323). This train of thought can be seen as a paraphrase of the Collingridge dilemma, which plays a crucial role in the next phase of the debate (see section "The Collingridge Dilemma").

Eventually, she reverses Nordmann's argument of opportunity costs. Even though some visions might be more or less "speculative" and so more or less probable, they can promise great goods worthy of consideration. The problem is to distinguish between "worthwhile, promising projects and unacceptably speculative projects" (Roache 2008, 325) and this is, as we have seen so far, a difficult task. She continues:

> On the one hand, we make use of cost-benefit analysis to decide which projects to pursue; and on the other, there seems to be a limit to the amount of improbability we are willing to tolerate, regardless of the value of the good we stand to secure if the dice roll in our favour. Since it is unclear where the division lies between worthwhile projects and unacceptably speculative projects, there exists the danger of a slippery slope, with increasing resources gradually being directed at increasingly unacceptable projects. (Roache 2008, 325)

She notices that many proposed solutions for this problem – such as Nordmann's – are present focused: new projects need to prove to society or politics that they are promising enough to neglect other present issues awaiting solutions. But, are society and politics prepared to undertake such assessment critically? The answer seems trivial in light of the views presented above: Of course, they are not the best at judging such projects because their opinions are "often influenced by fads, bias, propaganda, prejudice, and misconception" (Roache 2008, 325). Furthermore, the focus on present problems and worries is "short sighted": "[…] offering a solution to current problems when taking a more long-term view could either solve the current problems along with important problems that may arise in the future, or dispense with the need to address the current problems" (Roache 2008, 325–326).

As a current example, she mentions the problem of antibiotic resistances. The prescription of antibiotics was the right response to the diagnosis of an illness and did, therefore, meet the problems and concerns of that time. However, its excessive use was short-sighted because it gave rise to resistances. She concludes that we need to develop a kind of long-sighted perspective to reduce or, in some cases, prevent future problems from emerging:

> If ethicists are to be restricted to considering current and emerging technology, then allowing current social concerns to shape the direction of science is a short-sighted strategy that risks failing to emphasise the importance of taking a long-term approach. The best long-term approach sometimes requires allocating resources to projects that neither directly address any immediate need or problem, nor are known to have a high chance of success. (Roache 2008, 326)

It seems to me that she overlooks her own argument of biased credibility here. One can argue that penicillin revolutionized the treatment of bacterial infections, because it was new and more effective than other existing treatments. Thus, the knowledge about the long-term usage was little. How could the scientists and doctors know about the possibility of antibiotic-resistance in advance? The answer is that they could not have known until very recently. Furthermore, the long-term approach itself is bound to the present knowledge and beliefs: we can only think of things that have happened before or are imaginable within the set of actual experiences. It reminisces the "trends" Nordmann mentioned. There are trends and developments like antibiotic resistance which must be taken into account in our political and social actions, just as Nordmann suggests. It seems like the long-term approach is in line with Nordmann's conception of "trend influencing," hence his solution does incorporate a long-term perspective.

Despite biases in the perception of possibility and credibility, Roache highlights the importance of goal setting for the development and handling of technology and their implied values regardless of the different probabilities connected to those goals. Even though a vision might be improbable, society could decide to pursue its realization because of the great good it promises. According to her, ethicists should, therefore, concern themselves with "speculative" visions along with their implicit goals. Particularly, the goals and values need to be concretized so that visions can be judged regarding their potential to fulfil certain values:

> [I]f reaching a particular goal is important enough, we often judge it worthwhile to work towards it even if the chances of success look to be unpromising. However, arguably, there is in general too little discussion of which goals are the most important ones to aim at. […]

> Reflecting on where our most important values lie, and how we might work to maximise them, is surely an important step towards ensuring that ethical concern, and other valuable resources, are not squandered. (Roache 2008, 326)

As shown above, Armin Grunwald also uses the argument of biased credibility. He, too, criticizes Nordmann's argument of opportunity costs and his argument of influence but draws a different conclusion than Roache.[7] In general, he agrees with Nordmann that ethical enquiries of speculative visions must not be taken seriously for action recommendations and that we need a tool, vision assessment, to identify those purely speculative visions:

> [T]he simple possibility of thinking something cannot be employed as an argument or part of a chain of arguments, but only leads to aporias. If the future conditions that ethics reflects on were in a radical sense purely speculative, then the ethical advice that might follow from this reflection could not lead to anything of consequence. [...] [E]thics would not set limits on technology, but technology—as a result of its uncertain future—would set limits on ethics. [...] Thus the only result from the accurate diagnosis that there is high degree of speculation in parts of nanoethics is the demand that procedures be developed in order for us to be better able to judge the degree of speculation. (Grunwald 2010, 93–94)

In a first step, he criticizes Nordmann's argument of influence that stated speculative visions of a technology could hinder technological developments, even though it is very probable to create a desirable social outcome. Because of our limited knowledge and the implied bias "[t]his decision – here mere speculation and there justified concern – cannot be grounded" (Grunwald 2010, 95). Therefore, we should not dismiss ethical analysis of speculative visions *per se*. Secondly, he argues that the argument of opportunity costs phrased by Nordmann "is not tenable at all." Grunwald's defense is twofold:

> 1. Apparently literally following the term 'nanoethics', Nordmann/Rip consider the entire speculative debate about nanotechnology to belong to *applied ethics*. [...] Yet many of the studies about speculative aspects of nanotechnology can in no way be attributed to applied ethics if the participating persons, the arguments employed, and the literature used are taken into consideration. Many of these studies stem, to give several examples, from the philosophy of technology, anthropology, philosophy of mind, and the theory of artificial intelligence [...].

2. The persuasive power of the opportunity-cost argument depends fundamentally on the finiteness and constancy of the variable 'available resources'. The new visions of nanotechnology have attracted additional resources from parties interested in such issues (philosophy, STS). The pie has grown considerably. The opportunity-costs argument thus does not hold up. (Grunwald 2010, 95)

Grunwald notices that "nanoethics" "is a misleading umbrella term" (2010, 96): It does not only include applied ethics but also, for example, philosophy, anthropology, and STS. If "nanoethics" does not only give policy or action recommendation, the argument of opportunity costs collapses. To avoid this semantic ambiguity, he suggests the term "explorative nanophilosophy"[8] for reflection on visions of the future as its own sub-discipline.

The Collingridge Dilemma

Before presenting Grunwald's full concept of explorative nanotechnology, I should keep the earlier made promise and introduce the Collingridge dilemma. In Grunwald's account, it plays a major role and serves to differentiate explorative (nano)philosophy from (applied) ethics. Additionally, Nordmann (2010) dedicated a complete paper on it. As we will see, it also plays a key role in the plea for hermeneutic TA. The Collingridge dilemma was first articulated as following:

> [T]he social consequences of a technology cannot be predicted early in the life of the technology. By the time undesirable consequences are discovered, however, the technology is often so much part of the whole economic and social fabric that its control is extremely difficult. This is the dilemma of control. When change is easy, the need for it cannot be foreseen; when the need for change is apparent, change has become expensive, difficult and time consuming. This is the *dilemma of control.* (Collingridge 1980, 11, italics in original)

In the beginning of a technology's development, we can easily change, design, and control, even abandon the technology altogether. However, at this stage we do not know the social outcomes of it because there are too many uncertainties to properly assess whether the technology will harm society or provide great benefits. As there are no grounds of concern, but the possibility of missing out on its benefits, we should not engage in influencing or controlling the technology. If we wait for the technology to diffuse in society so that we can assess its impact, we are too late to control it – it has become an integral part of society and is tightly interwoven with human action as a socio-technical system. In the end, society and politics have to adapt to the new technology and take up a purely reactive role (Collingridge 1980, 16–18). In this way, the Collingridge dilemma is a cornerstone of the previously sketched debate. Grunwald states this explicitly: "Nordmann/Rip's criticism can be read as a diagnosis that nanoethics, in order to avoid the Scylla of the Collingridge dilemma (namely of coming to late), must fall victim to the Charybdis (i.e. inevitably being too speculative)" (Grunwald 2010, 96).

Explorative Nanophilosophy

In contrast to the gravity that the original outline of the dilemma by Collingridge suggests, Grunwald is not convinced that the dilemma is as concerning for the methodology and process of ethical evaluation as it seems at first glance:

> The Collingridge dilemma is exaggerated. The question of whether ethics should start early or late, whether it should be prospective or only start when reliable statements about consequences are available poses a false alternative. The issue here is not an either-or but a differentiation of ethical reflection in line with the problem at hand and with the validity of the available knowledge of the consequences. Ethical reflection differs conceptionally and methodically depending on whether it is concerned with measurable consequences of technology or with just imagined or even merely speculative ones. Above all, it also serves *different purposes*. (Grunwald 2010, 96, italics in original)

Grunwald makes a clear distinction between applied ethics which concerns itself with "measurable consequences of technology" and explorative nanophilosophy which concerns itself with "imagined or even merely speculative ones." Each serves different functions and purposes. Whereas applied ethics' main purpose is to provide recommendations for actions by judging measurable consequences ethically, explorative nanophilosophy has a variety of different aims. By analyzing speculative visions, it can lay the foundation for future debates in applied ethics and helps us "identifying knowledge gaps" (Grunwald 2010, 97), so it takes on an agenda-setting-function. Finally, we can ask – as Nordmann already suggested – what those visions tell us about us today:

> [W]hat is going on in the speculative developments being considered, what is at issue, which rights might possibly be compromised, which images of man, nature and technology are formed and how they change, which anthropological issues are involved, and which designs for society are implied in the projections for the future. (Grunwald 2010, 97)

These questions and issues are of "conceptual, *hermeneutic*, philosophical (philosophy of technology) or anthropological" (Grunwald 2010, 97, my italics) nature, and so we can distinguish three different fields of explorative nanophilosophy.

In the following, I will briefly present all three fields with their respective concerns with a focus on the hermeneutic component.

The Hermeneutical Turn

This is the first time in the discussion that the term "hermeneutics" is invoked. First, the promises of nanotechnology call for an analysis of the kind of knowledge it claims to produce and the premises behind the knowledge. Is the idea of the world as suggested by nanotechnology coherent? Second, in the discourse of nanotechnology and human enhancement, we have to ask what anthropologies are implied by the different visions. What is a "human," and what is the difference between men and machine (if there is any)? The third and last field touches the hermeneutic-philosophical dimension of nanotechnology. Two main questions are raised: How does nanotechnology change the abilities of human control and invention? Which function does (the vision of) nanotechnology play in present debates ("*cipher* of the future" [Grunwald 2010, 99, my italics])?[9] Particularly the latter question is of great interest for the further debate of visions of the future. Explorative nanophilosophy gains in this way a hermeneutic task, i.e. the deciphering of the visions' content as visions used in debates. Grunwald summarizes:

> It is more to understand what's at stake and issue, a contribution to a 'hermeneutics' of possibly changing elemente [sic] of *condition humana*. In this way, explorative philosophy can prepare the groundwork for applied ethics and for the technology assessment of the developments when they become more concrete. (Grunwald 2010, 99, italics in original)

There is a remaining concern: How should explorative (nano)philosophy generate knowledge? As demonstrated so far, visions of the future are often speculative, but there are ways to deal with its speculative nature. Vision assessment is one of them:

> Explorative philosophy must, in order to circumvent this criticism, not be epistemologically blind but must develop methods and procedures of assessing pictures of uncertain futures with respect to their degree of rationality. It needs clarification of the cognitive and normative content of the partially speculative future conditions and consequences of nanotechnology. […] Any epistemology of futures is a kind of 'deconstruction': disassembling the elements of those

> futures such as pieces of knowledge, uncertainties, ad hoc assumptions, values and other elements as well as the ways of their 'composition' into one picture of the future, assessing the validity or plausibility of those elements and then assessing the overall 'rationality' of the considered picture of the future compared to the rationality of others. (Grunwald 2010, 99–100)

It is the first time in the discussion that the meaning and the implications of the use of the term "future," as often employed by TA scholars, are made explicit. Every vision and prediction is to a certain extent uncertain and perhaps speculative because the future is unobservable for us; it is something that did not happen yet, and many things might happen which we do not (and cannot) take into account when predicting. The future is, thus, first and foremost present as a form of imagination

> *created* by humans, individually or in teams [...]. Each future we are talking about is a *construction*: an extrapolation of current trends, a vision promoted by scientists, a scenario developed for the purposes e.g. of urban planning or in the climate change context, or a prediction of demographic change. (Grunwald 2010, 99, italics in original)

This insight seems "trivial" (Grunwald 2010, 100), but as we have noticed, Jasanoff's critique of the "technology of hubris" is written implicitly to reveal this notion of the future, and Nordmann's critique articulates this clearly, too. Nordmann's diagnosis of the "if and then"-syndrome attacks the treatment of visions of the future as if they are inevitable. As trivial as it seems, the consequences of this view are quite severe for our way of dealing with uncertainties and the future:

> At first it makes clear, that futures (and that is the reason why the plural 'futures' should be used) are inevitably part of the respective present time in which they are constructed and communicated. Secondly, it demonstrates that we do not have any direct access to the future as it really will be in due time. We only can think about futures according to our present-day's knowledge, our values, fears and expectations—and according to our way of 'constructing' pictures of the future. This leads to the third aspect: the way how futures are constructed and which ingredients are used [...] is decisive for their content. Finally, this implies that any assessment of the 'quality' of futures—whatever this would mean in detail—has to consider the construction process and the ingredients used. It does not make sense to use the probability of matching real future developments by present images of the future as a quality criterion because this cannot be made operable. If we want to compare

> different, possibly diverging futures we have to look at them as constructions and to assess the quality of the construction and its elements. (Grunwald 2010, 100)

In short, "the future" is part of our respective present. We cannot transcend our present perspective, because it is everything we can possibly know. The future as envisioned is nothing more than an expression of our knowledge and emotions in the present. If we are to assess visions of the future, we have to acknowledge them as constructions and deconstruct them, so that the different components can be judged separately and the construction as a whole with regard to the context of its creation. Only then can we seriously compare different visions of the future without making any claims about their probability. Present trends and the envisioned scenario in the future are categorically different as discussed/shown before. In this way, nanoethics or rather explorative philosophy liberates itself from its task as policy advisor and assesses visions of the future in order to prepare ethics for future debates, to reveal cognitive and normative elements that require further investigation or discussion, and lastly to learn something about us today.

Forensics of Wishing

In the same year in which Grunwald published *From Speculative Nanoethics to Explorative Philosophy of Nanotechnology*, Alfred Nordmann (2010) also made comments on the de-problematization of the Collingridge dilemma similar to Grunwald's and dismissed the dilemma as such. He discusses the function of TA and criticizes its aims and premises:

> The very fact that the Collingridge dilemma is considered a dilemma that is in need of a solution shows that TA is not just implicated in the age of technoscience, but actually participates in its spirit of technoscience. […] In particular, the project [of technosciences] is one of exceeding or outflanking the limited temporal horizon of our sphere of influence, but it is fairly obvious that we dispose of occurrences in the present, at best, and that we have no control over events that have occurred already in the past or that will occur in the future. To consider this as a dilemma is tantamount to viewing the present as an obstacle that can and needs to be overcome. (Nordmann 2010, 9–10)

As we can see, he draws the same conclusions concerning the use of the word "future," but he criticizes TA instead of applied ethics. The discussion about visions of the future has, thus, transformed into a critique of TA as a predictive method, as a mode of constructing and assessing visions. According to Nordmann, TA as a technoscience aims to exceed "the limited temporal horizon of our sphere of influence." By presenting scenarios or consequences of technologies, TA expresses a pathos of shaping or designing the future because it is "accessible" to us today. Nordmann calls this observation "the transgressive hubris of the technosciences" (2010, 10). This observation resonates well with Jasanoff's critique of the "technologies of hubris" discussed earlier. In general, TA promises to provide accurate and reliable visions of future consequences without acknowledging its own limitations and the involved uncertainties. Nordmann elaborates on an intuitive counter-argument for the designability of the future: The future is always a product of past and present actions, so we should be able to design it. Therefore, we should try to design the future by acting in a certain way at a certain time. Only then can we think of the Collingridge dilemma as a problem of control which can actually be solved (Nordmann 2010, 11). The aspirations of promoters of human enhancement clearly demonstrate this: We will be different people in the future because we will have invented and altered technologies, which also leads to a change of us (see the discussion of "Conflation" earlier). Consequently, we should try to shape us according to our will, but

> [t]here remains a categorical difference between creating the conditions in which the future will take place and steering, shaping or controlling what the future will be like. The fact that I will become a different person in a different technological environment does not mean that I can design myself by designing new technologies. [...] Though we do, in fact, produce the future, the future remains utterly contingent and is not, therefore, something that is made according to our wish or will and is not a target of socio-technical design. (Nordmann 2010, 10)

One might question why we should commit to the future given its inaccessibility. In the previous discussion, "future" was only used in terms of a moment in time, which has not come yet, but we can easily see that the term "future" is ambiguous. The discussion about scientific knowledge, for instance, needs the "future" as a temporal reference when we will obtain truth (Nordmann 2010, 13), which might never be reached. In this sense, the future is more an aspiration of dedication for self-criticism and progress. Therefore, the meaning of "future" in the technosciences and TA, respectively, can be revealed:

> [W]hen the technosciences speak of shaping the future, they are not referring to the future of humanity and society, but rather to the realisation of a potential or the fulfilment of a wish, that is, they are talking about a future that is fully contained in the present. And if to posit a potential or to formulate a wish is the same as shaping the

> future, TA needs to be a forensics of wishing and can thus engage the future without going beyond the present. [...] Moreover, even though in most cases it takes time for a wish to be fulfilled, the fulfilment does not lie in the future as much as it lies in the wish. The wishes themselves are not just the products of technological development, but also its drivers. To the extent that they inform technological development, they are suitable subjects for TA—they hold the image of a technologically conceived future that is fully contained in the present. (Nordmann 2010, 12)

The future of technology lies, according to Nordmann, within "the fulfilment of a wish," of reaching a certain technological potential or function. It is the wish of being able to solve certain problems with the help of technology or of enabling certain technological features.[10] But, the wishes themselves are not fixed, they can be rephrased or abandoned, and most importantly, they are part of the present and have an impact on current debates or decisions. This is the reason why we should assess them in a "forensics of wishing" as a kind of vision assessment which could sketch the path from a wish to its fulfilment. However, such forensics could also entail an analysis of the "idea of fulfilment that is contained in the wish," and so this assessment can go without an estimation of "probability and credibility" (Nordmann 2010, 13). The objects suitable for such forensics of wishing are not only texts but also simulations or prototypes (see Dickel in the present volume), which means that the institution of TA itself can become an object of a TA assessment. Hence, his proposition expands the possibilities and aims of vision assessment as introduced by Grin and Grunwald in 2010.

From Vision Assessment to Visioneering Assessment

In 2012, we first encounter the notion of "hermeneutic TA" in Grunwald's epilogue of his earlier mentioned book *Technikzukünfte* (Technofutures). The last section of his book expands earlier thoughts about hermeneutics expressed in "explorative nanophilosophy" and concretizes the meaning of "hermeneutics" in the context of TA. First, a hermeneutical analysis can only be of linguistic nature because language is the main medium to express thoughts about the future. However, "language" covers various media of communication including "classical" languages like English to more scientific and perhaps more formal, mathematical languages. Such an analysis is concerned with the visions of the future expressed in texts, debates, regardless of their context of use, but also diagrams, pictures, or scenario-modeling often used in TA.[11] Second, a hermeneutical TA has the task to make scientific jargon comprehensible to the public in the form of reconstructing "if-then"-statements. "Pure" texts and debates, on the other hand, need another kind of analysis: Hermeneutic TA should be more concerned about a cultural-linguistic analysis which uncovers the cultural-historic background of visions together with their implicit meanings. It should deliver explanations of visions and their following debates as well as examinations of the anthropological, philosophical or sociological questions raised. Eventually, such a hermeneutic TA should enable well-informed debates or prepare future debates about technology (Grunwald 2012, 283–285). The tasks and possibilities of vision assessment as a tool of TA are once more expanded in favor of a more hermeneutic approach. By now, it should be clear that such an analysis has to be an interdisciplinary effort instead of one for individual philosophers and ethicists.

This approach is not the end of our journey to hermeneutic TA though, because the notion of hermeneutic TA was (and still is) evolving and expanding. One crucial step in our reconstruction is missing toward which the gradual widening of vision assessment points; the development of "visioneering assessment." The discussion so far has demonstrated that vision assessment deals with the visions themselves: What are the cognitive and normative elements? Is the overall construction of a vision with the identified elements "rational"? What are the hidden premises and inexplicit norms of the visions? How are visions used in public debate? Where do the visions originate from – culturally or historically? What do visions tell us about us today? What are the differences between the diverging visions? The former four questions represent a "epistemological 'deconstruction'" (Grunwald 2013, 31), whereas the latter four questions represent a "hermeneutical construction" (33). The vision itself is analyzed in great detail, but little is said about the *process of constructing*, which cannot be neglected:

> Techno-visionary futures do not exist per se, nor do they arise of their own accord. On the contrary, they are 'made' and socially constructed in a more or less complex manner. Futures – be they forecasts, scenarios, plans, programmes, visions, speculative fears or expectations – are 'produced' using a whole range of ingredients such as available knowledge, value judgements and suppositions. [...] Visionary futures are frequently created by scientists and science managers who at the same time are stakeholders with their own interests. One possible scenario is that visionary futures suggested by science could dominate social debates by determining their frames of reference [...]. (Grunwald 2013, 29)

Visioneering

In 2013, when Grunwald wrote about the product-character of visions, historian Patrick McCray dedicated a whole book to the process of visioning and engineering. His observation of many modern futurists underscore that they are in fact visionaries and engineers at the same time due to their ability to use their "engineering and technical knowledge" (McCray 2013, 10) in order to create visions of the future and contribute to their realization. To properly describe such individuals, groups, institutions etc. and their activities, he proposes the terms "*visioneers*" and "*visioneering*" (10, 13, italics in original) for the process itself. Visioneers develop or rather construct a vision of the future and try to attract attention so that their projects are funded. They are convinced that their vision is worth pursuing and has to be realized to provide benefits or to solve current problems:

> [*V*]*isioneering* means developing a broad and comprehensive vision for how the future might be radically changed by technology, doing research and engineering to advance this vision, and promoting one's ideas to the public and policy makers in the hopes of generating attention and perhaps even realization. [...] The visioneers' hybrid nature—a combination of futurist, researcher, and promoter—and the influence they sometimes attain compels us to consider how they interact with other actors in broader systems of technological innovation. (McCray 2013, 13–14, italics in original)

McCray underlines three main visioneering efforts: visioning, engineering, and promoting. Especially the latter is important for TA. The aim of influencing the public opinion and debates to increase their visions' traction and support has actual, present-day impact. In the same issue of *Science, Technology and Innovation Studies* in which Grunwald published *Techno-visionary Science*, Alfred Nordmann summarizes the visioneers' goals and mindset as follows:

> What visioneering aims for is to exhibit a compelling causal link between a state A (technological work-in-progress) and a state B (a future so desirable as to mandate its realization) such that A will actually lead to B while B necessitates A. (Nordmann 2013b, 89)

The Call for Visioneering Assessment

In this way, a "visionary narrative" is framed and promoted which leads to the "marginalisation of alternatives" (Coenen and Simakova 2013, 15) and a "tunnel vision" (Nordmann 2013b, 90). By suggesting that their vision is the only one which will lead to the desirable outcomes, "the future" seems to be closed down by visioneers. Thus, they can – or rather want – to set the political agenda as well as the agenda of TA by attributing immense priority to their visions through promises of high desirability and possibility of realization (Nordmann 2013b, 90; Grunwald

2013, 29). Although it remains inexplicit in his own writings, Nordmann recognizes that Grunwald suggests a shift from vision assessment toward visioneering assessment:

> Now, he [Grunwald] is no longer looking at visions as representations of an emerging technology but is shifting attention from the emerging technology to quite another technology, namely to the institution of the causal relations between A (technological work currently in-progress) and B (a desirable future that will be produced by this technological development). Visioneering assessment looks at the engineering process that has produced a compelling vision of a technological future. It thereby not only determines whether we really have to take this compelling vision seriously, but also opens the black box of umbrella terms by exposing our struggles over visions of the future. (Nordmann 2013b, 93)

Consequently, visioneering assessment has the task to uncover the process in which visions are constructed. Visioneering as a special form of engineering produces causal links which have to be uncovered, and the inevitability of this link has to be challenged. To do so, visioneering assessment has two tools as introduced by Grunwald and Nordmann, respectively. As mentioned before, visions go through a complex and yet opaque process of defining, redefining, and abandoning, and then just a few of those selected visions get the attention of mass media and eventually policy makers. The first tool is, therefore, an analysis of the "biography" (Grunwald 2013, 30) of the visions along with the factors and reasons for their selection: team or company, mass media, decision-making, with each step increasing its impact on public opinion and agenda-setting. In Grunwald's words:

> The main objective would be to generate more knowledge about and greater insights into the social processes surrounding visionary futures, from their emergence and dissemination via different communication channels to their possible impact on decision-making in the policy arena and other arenas of public communication and debate. (Grunwald 2013, 31)

It is important to know which factors led to the selection or abandonment in order to judge whether the selection process was "successful" in identifying the most desirable and "rational" vision. To complete the tasks of visioneering assessment, Nordmann suggests the analysis of "past debates about techno-visionary futures":

> This is of interest in its own right since it challenges contemporary visioneers to detail why they believe that the realization of a perennial

> dream is finally becoming a real possibility just now. While this qualifies as well as sharpens the techno-visionary futures under consideration, it does not yet open the space for 'thinking in alternatives.' This can be done however, when one steps away even further from the particular visioneering construction by considering not just its genealogy but also the reception of its previous incarnations. It is those who in the past rejected certain techno-visionary ideas who are most likely to point us to alternative conceptions of technological and human development. (Nordmann 2013b, 93)

Dismissed, earlier envisioned futures become problematized again to challenge the visioneer to defend his claims about the realizability and desirability of his pursued vision. On the other hand, we can look at similar visions of the envisioned technology which were rejected in the past. These past visions were constructed through different cognitive and normative elements. Their deconstruction could uncover other divergent ideas of technological feasibility and desirability, which in turn challenge the compelling causal links mentioned above. The space of *envisioned* futures, thus, opens up again. In conclusion, visioneering assessment analyses the construction process of visions, makes hidden assumptions and implicit norms explicit and questions the "tunnel vision[s]" that are by-products of visioneering pursuits. Vision assessment focuses on the components and the overall composition of a vision itself, whereas visioneering assessment illuminates the reasons for the concrete composition of a vision along with a critical evaluation of the causal links between desirability and realizability; in this way, vision and visioneering assessment complement each other. Through such detailed analysis, they enable open and informed public deliberation about visions of the future.

Recapitulation

Before discussing hermeneutic TA in general, I want to consider one last discussion paper published by Nordmann (2014) called "Responsible Innovation, the Art and Craft of Anticipation." Together with its responses, it provides the perfect backdrop to summarize the results and arguments of the discourse about visions of the future. Nordmann discusses in this paper the three dimensions of "anticipation": anticipating "the unexpected in the world as we know it," anticipating in the sense of "[g]etting ahead of ourselves" and anticipating for a "politics of the future" (2014, 88–91). The first meaning of anticipation is linked to the discussion about the argument of biased credibility: We can only prepare for things that have happened before or that are imaginable because of previous experiences. The second one is the notion of future ("non-trivial future" [Nordmann 2014, 90]), which has been dismissed earlier as being inaccessible. He puts forward two arguments for the dismissal of that notion of future, an old and a new one. First, there is the argument of our bias toward the present (see section "Argument of Biased Credibility"). The point departs from our inability to transcend our present knowledge because of the limited set of experiences available to us. This leads to a second argument of the inaccessibility of a non-trivial future that promises "scientific and technological breakthroughs" (Nordmann 2014, 92) which are not part of present trends like climate change.[12] Through this observation, he criticizes the use of the term "emerging technologies." As he puts it:

> [A] metaphor [...] which suggests that the emergence even of radically new technological capabilities can be linked to a rising tide – it will just go on and on, and at some point the dams to a new world with new capabilities, opportunities, and risks will simply break. (Nordmann 2014, 92)

In a response in the same issue, Cynthia Selin (2014) develops three points of criticism of Nordmann's depiction of non-trivial futures, which shall be addressed in turn. First, she seems to believe that the leap in time we have to take to get a certain technology is "short": "The average time horizon for most policy relevant futures work that Nordmann appears to be concerned about it is about 10–15 years, a time in which one could question how 'profoundly different' 'the world' might actually be" (Selin 2014, 105).

According to her, the majority of decisions we are talking about today concerns the next two decades. It is, therefore, questionable that the world will be essentially different from the present. She is right that those decisions concern this "short" time span, insofar as we believe the sometimes, prophecy-like predictions to be true.[13] Indeed, the things could happen in the near future, but they also could not. That is exactly the point: we cannot really know if and when such a "breakthrough" might occur, and consequently we should not pretend to be able to address actual solutions or problems in this non-trivial future with our decisions now. Second, she suggests that "Nordmann seems to be caught in a conventional, linear approach to thinking through time, one divorces from what social theorists have been learning about temporality" (Selin 2014, 105).[14] To me, this critique seems to stem from an omission in Nordmann's paper: he does in other writings and personal conversations acknowledge a discontinuous conception of time (see Proactionary-Precautionary-Debate, Chapter 8). Selin's third argument charges Nordmann with a fallacy:

> Nordmann also seems to present a logical fallacy. If our values change with changing technologies, then are we not as likely to misunderstand the past as the future? If we are so changed that our

> future-oriented imagination fails us, would not the same be true of our historical imagination? (Selin 2014, 105–106)

This is an important and apt observation: Nordmann's approach does indeed not differentiate between a non-trivial future and a distant past (see also Sand 2018). Because those points in time are not directly accessible; they are only indirectly accessible (ibid.), *interpretative* or rather *hermeneutic* work is necessary in each case. As James Wilsdon points out correctly, "[h]istorians are good at judging the interests that lie behind differing interpretations of the past, as well as exploring their validity" (2014, 110). The past can, just like the future, never be accessed directly. Knowledge about it can be obtained only through inference of what we currently consider traces thereof. Other than the future, which does not yet exist, the past does not exist anymore and, thus, presents itself merely as a construction. That does not contradict Nordmann's approach, on the contrary, it stresses even more the need for a hermeneutic approach.[15]

Coming back to Nordmann, he proposes to only take present trends as extrapolation of current data seriously for action recommendations, and a hermeneutic analysis of visions as visions, as "cipher[s] of the future":

> [T]he future influences the present by providing a standard or measure of progress [...] because the future represents regulative ideas of a better world in which welfare and justice might be achieved. [...] [N]ormative conceptions of a better world can orient evaluative discourse in a direction of change only if the future is not treated as an object of anticipation and not conceived as what will or might come to be. Only then scenario thinking becomes liberated and can serve a thought experimental heuristic, allowing it to break the mold of preparing us for what is presumably already in the making and half-destined to be, and allowing us instead to envision a better world and thus to judge scenarios for their desirability, unencumbered by insinuations of inevitability. [...] *Historical sensitivity* also allows us to engage in a *hermeneutics* of envisioned futures, that is, an interpretation and critique of envisioned alternative worlds as they appear in the writing of promoters and critics of current technological research. (Nordmann 2014, 91–93, my italics)

The practical benefit of such liberation of the future is that it becomes a reference point for comparable progress, like in scientific enquiry. Our present notions of justice and welfare can converge in utopian thinking, so that they can serve as a guideline for political deliberation today. It is, thus, crucial to make hermeneutic analyses of visions in order to contextualize them in their respective time of origin: only the circumstances of its construction process – in social, historic and cultural perspective – can acknowledge visions of the future as one vision among many other possible visions. In a response to Nordmann, Simone van der Burg alludes to Paul Ricoeur's insights:

> A movement can only qualify as an action, in relation to other actions that follow it in time, which are carried out by the actor him or herself or by other people in a practice. But if future actions convey meaning to present actions, it is important to find out in relation to what sequence of follow-up actions it should be understood, or to what practice it belongs. [...] [T]hey [future imaginations] help to give meaning to present research and development actions, based on imaginations of the actions that could follow them. [...] Imagining the future is more like an experiment that helps to understand the meaning of the instrument that researchers envision to make, as well as the significance of some of their research actions that contribute to creating it. [...] The reason to engage in such imaginations is to inspire thoughts about what researchers would want their technology to be, and to throw a different light on the significance of specific research actions. (van der Burg 2014, 101)

Hermeneutic TA – A First Outlook

We have provisionally reached the destination of our journey. I reconstructed the long way to hermeneutic TA as a dialogue between Armin Grunwald, Alfred Nordmann, and other authors, who have challenged their conceptions of the future. This dialogue started from a critique of the handling of visions of the future in the discourse of nanotechnology. The reason for the engagement in discussion of visions of the future is their real impact on agenda-setting, public opinion and decision-making. Crucial for the initial critique and the following discussion were the arguments of opportunity costs, influence, and biased credibility. In the course of this discussion, the notion of the future liberated itself from a kind of technological determinism. The next station was vision assessment, which was the first hermeneutic tool added to the repertoire of TA that was complemented by visioneering assessment. In this last section, I will outline one possible meaning of hermeneutic TA.

As I demonstrated above, in the beginning hermeneutic TA was just a generic term for all the different hermeneutic methods in TA. Explicitly, there are vision and visioneering assessment as two dimensions of this concept. Their main aims are to open up the space of alternative futures, in contrast to the closing character of promoted visions presenting us with a fait accompli. Hence, hermeneutic TA also aims to enable well-informed public deliberation and decision-making. Objects of interests are visions expressed in the public along with scientific writings and speeches, including pictures, diagrams, and models. Grunwald (in the present volume) proposes, for example, how a hermeneutic analysis of models could be put into practice. Nordmann and Grunwald (in the present volume) present case studies of how hermeneutic TA could be put into practice. Also, real technological artefacts like prototypes (see Chapter 9) are intended/sought to be analyzed because of the implicit messages inscribed in them. As a consequence, the range of objects eligible for hermeneutic analysis expands even further. Furthermore, the process of visioneering must be reconstructed and critically judged. TA itself as a process in which visions of the future are made becomes in this way through and through an object to TA. Hermeneutic TA grows thus into a self-critical instrument, which emphasizes the limitations of "traditional" TA. All these measurements have the purpose to contextualize different aspects of a vision from various angles in order to learn something about us today and most importantly to demonstrate that each vision is just a possible vision among many other alternative visions. On the one hand, hermeneutic TA increases our need for orientation due to the number of alternatives that "emerge," but at the same time it helps us to judge each vision's desirability and rank them in a "rational" order.

This notion of hermeneutic TA is almost exclusively reactive though and thus not really satisfying. Hermeneutic TA reacts to the proposed visions by deconstructing and reconstructing them in order to question their implicit premises and (explicit) promises so that it becomes apparent that they are just possible visions among many others. In 2019, Grunwald responded to this shortcoming: It seems obvious that hermeneutic TA should take a more active, perhaps more leading role in "shaping" technology in the present. TA as a predictive method was engaged in scenario-building and vision construction ever since to enable informed and "rational" public deliberation. Keeping this role is sought with a sense of hermeneutic self-awareness. In other words, aside from the reflection on visions of the future, a hermeneutic TA should actively construct alternative visions of the future to increase the numbers of visions discussed in public deliberation (Grunwald 2019). To conclude, the aim or the promise of hermeneutic TA is clearly expressed in the title of his paper: *Shaping the Present by Creating and Reflecting Futures*. We do not shape the future itself but the present – guided by the (created) visions of the future and the discussions about them. To be clear, we do not have many explicit tools for creating visions systematically yet that fulfil the self-set standards of variety, but hermeneutic TA provides a framework in which such tools can indeed evolve and be put into practice (see e.g. Chapter 10). In the end, the hermeneutic tools for TA grew into a hermeneutic work mode of TA. In this way, hermeneutic TA is a newly constructed "technology of humility" that Sheila Jasanoff called for in 2003 to contrast the hubris connected to TA. Hermeneutic TA is in many respects "humble": It limits itself to the indisputable knowledge available at the moment and to influencing the present, not the (distant) future. Moreover, it recognized the limits of understanding socio-technical change by individual disciplines; hermeneutic TA can only be successful if exercised inter- and transdisciplinary.

Although it seems as if we have reached a conceptual endpoint in the journey to hermeneutic TA, the concept is still methodically challenged, expanded, and refined. Noteworthy amongst the recent engagements with hermeneutic TA are Martin Sand's publications (Sand 2019, reprinted in the present volume), in which he identified a blind spot or rather a chance for improvement in hermeneutic TA methodology. His main finding, taking Jasanoff (2016) and McCray (2013) as a starting point, is that many – if not all – of the most prominently discussed visions in TA were constructed and promoted by elite individuals or groups with great political influence on media and decision makers. Here, hermeneutic TA falls short to its own aims, i.e. opening the space of alternative visions. More "'profane' futures" (Sand 2019, 104) concerning unemployment, for instance, are excluded from the considered visions of the future. This observation points toward a call for the additional analysis of power-relations within the framework of hermeneutic TA. At the same time, it emphasizes the necessity to include and discuss visions of the future that may not at all be connected to any particular technology. As paradoxical as it seems for an institution with this name, TA must consider enabling non-technological futures. Christopher Groves (see Chapter 7) expands on this strand of idea and explores how those more "profane" futures could be implemented into the visioneering process.

It is obvious that hermeneutic TA is still evolving and self-criticizing in order to improve on providing lasting, democratic, and humble information about visions of the future. It is unclear which developments hermeneutic TA will undergo and which tools will evolve from its intellectual ground, but the following chapters of the book give a first impression of the future path, which hermeneutic TA might take, by demonstrating what TA can do if it does not take the technological *Engführung* of the future.

Acknowledgements

I would like to express my deep gratitude to Alfred Nordmann, Armin Grunwald, and Martin Sand, the editors of this book, for their encouragement and helpful feedback.

Notes

1 More precisely, she is in favor of alternatively implied visions of the future because our tools to predict such can only give us one picture of an eventually possible future among many others.

2 "Falsely" in this context means that on first glance something seems desirable but after meticulous analysis turns out not to be.
3 For a further discussion of social and technological determinism see also Grunwald (2012, part one).
4 Apparently, Grunwald refers here to the concept of "rationality" expressed in the theory of deliberative democracy by Jürgen Habermas (e.g. 1996).
5 In 2013, Nordmann published a paper on the use of "plausible" in connection with envisioned worlds. For this purpose, he analyzed the different meaning of the words "possible," "plausible," "feasible," and "probable" and the different occasions in which one can or should use them (Nordmann 2013a). Overall, he concludes that we cannot speak of plausibility of future worlds as we do in the present, because the world and the subjects inhabiting it will be different. As a consequence he suggests to use the term "plausibility squared" (128). In this way, we can acknowledge that the plausibility of possible future worlds itself depends on other claims about plausibility of trends as well as deductions of causal links in the present and how this eventually leads to a envisioned future. Therefore, we have to engage in a kind of hermeneutics of envisioned futures in order to obtain a clear picture of the claims of plausibility (128–131).
6 In this case, Nordmann writes that we need vision assessment of and debates about visions of technology to "bring out the full potential of technology" (2007b, 44). Ethicists should, in this sense, guide scientists and engineers.
7 Notably, he refers to Roache's approach, but states that his approach is intended to go beyond her's (Grunwald 2010, 99).
8 It is important to mention that Grunwald refers here directly to the field of nanotechnology as an example. One can apply his thoughts easily to any other field of technology which involves many speculative visions such as artificial intelligence or synthetic biology.
9 Grunwald expressed this thoughts about visions of the future and the function of them in the case of nanotechnology as early as 2006 in a contribution called *Nanotechnologie als Chiffre der Zukunft* (Nanotechnology as a Cipher of the Future), but he firstly expressed the idea for a explorative (nano)philosophy as a consequence of this perception of the future in this paper.
10 This notion of the future challenges the primary use of the ethical theory of consequentialism which is expressed formidably in the aims of TA. If we acknowledge the limitations of TA to predict consequences of decisions concerning technology development, we are forced to reconsider consequentialism as our main ethical theory for evaluation: hermeneutic TA proposes thus the additional consideration of the other two main ethical theories, namely ethics of virtue and ethics of duty.
11 For a further discussion of the role of modeling in TA see Chapter 10.
12 Even if we were to consider Moore's law to be correct and, thus, part of the present, the increasing complexity of technical systems would not automatically lead to technological breakthroughs. These breakthroughs are indeed not part of our present but would lie in the future.
13 Many of those prophetic predictions like Kurzweil's prediction of the emergence of strong AI turned out be wrong (Sand and Schneider 2017, 19).
14 She might be referring to concepts like "*chronos*" and "*kairos*" as presented by Steve Fuller in 2017: "In *chronos*, genealogical succession drives the narrative flow, with revolutions providing temporary ruptures which are quickly repaired to resume the flow. [...] In contrast, in *kairos*, there are recurrent figures who constitute the narrative but no default narrative flow, as the world

order is potentially created anew from moment to moment" (Fuller, 2017, 155–156, italics in original).

15 Grunwald actually compares the interpretation of texts or artefacts, which shall represent the future, with archaeology because those ancient artefacts are nothing more than ciphers from the past waiting for interpretative work (in conversation, December 8, 2020).

References

Amato, I. 1999. *Nanotechnology: Shaping the World Atom by Atom*. Washington: National Science and Technology Council.

Coenen, C., and E. Simakova. 2013. "STS Policy Interactions, Technology Assessment and the Governance of Technovisionary Sciences." *Science, Technology and Innovation Studies* 9 (2): 3–20.

Collingridge, D. 1980. *The Social Control of Technology*. London: Frances Pinter.

Dickel, S., and H.-F. Schrape. 2015. "Dezentralisierung, Demokratisierung, Emanzipation: Zur Architektur des digitalen Technikutopismus." *Leviathan* 43 (3): 442–463.

Fuller, S. 2017. "The Posthuman and the Transhuman as Alternative Mappings of the Space of Political Possibilities." *Journal of Posthuman Studies* 1 (2): 151–165.

Grin, J., and A. Grunwald, eds. 2000. *Vision Assessment: Shaping Technology in 21st Century Society: Towards a Repertoire for Technology Assessment*. Heidelberg: Springer.

Grunwald, A. 2006. "Nanotechnologie als Chiffre der Zukunft." In *Nanotechnologien im Kontext*, edited by A. Nordmann, J. Schummer, and A. Schwarz, 49–80. Berlin: Akademische Gesellschaft.

Grunwald, A. 2007. *Ethical Inquiry Meets Future Projections: The Case of Human Enhancement*. Karlsruhe: ITAS Pre-Print.

Grunwald, A. 2010. "From Speculative Nanoethics to Explorative Philosophy of Nanotechnology." *NanoEthics* 4 (2): 91–101.

Grunwald, A. 2012. *Technikzukünfte als Medium von Zukunftsdebatten*. Karlsruhe: KIT Scientific Publishing.

Grunwald, A. 2013. "Techno-Visionary Science: Challenges to Policy Advice." *Science, Technology and Innovation Studies* 9 (2): 21–38.

Grunwald, A. 2019. "Shaping the Present by Creating and Reflecting Futures." In *Socio-Technical Futures Shaping the Present: Empirical Examples and Analytical Challenge*, edited by A. Lösch, A. Grunwald, M. Meister, and I. Schulz-Schaeffer, 17–35. Wiesbaden: Springer.

Habermas, J. 1996. *Die Einbeziehung des Anderen: Studien zur politischen Theorie*. Frankfurt am Main: Suhrkamp.

Jasanoff, S. 2003. "Technologies of Humility: Citizen Participation in Governing Science." *Minerva* 41: 223–244.

Jasanoff, S. 2016. *The Ethics of Invention: Technology and the Human Future*. New York: W.W. Norton.

Johnson, D. G. 2007. "Ethics and Technology 'in the Making': An Essay on the Challenge of Nanoethics." *NanoEthics* 1 (1): 21–30.

McCray, P. 2013. *The Visioneers: How a Group of Elite Scientists Pursued Space Colonies, Nanotechnologies, and a Limitless Future*. Princeton: Princeton University Press.

Nordmann, A. 2007a. "Gestaltungsspielräume in der Nanowelt: Eine Space-Odyssee." In *Zukunftspotentiale der Nanotechnologien: Erwartungen, Anwendungen, Auswirkungen*, edited by D. Korczak and A. Lerf, 159–184. Kröning: Asanger.

Nordmann, A. 2007b. "If and Then: A Critique of Speculative NanoEthics." *NanoEthics* 1 (1): 31–46.

Nordmann, A. 2010. "A Forensics of Wishing: Technology Assessment in the Age of Technoscience." *Poeisis Prax* 7: 5–15.

Nordmann, A. 2013a. "(Im)Plausibility^2." *International Journal Foresight and Innovation Policy* 9 (2/3/4): 125–132.

Nordmann, A. 2013b. "Visioneering Assessment: On the Construction of Tunnel Visions for Technovisionary Research and Policy." *Science, Technology and Innovation Studies* 9 (2): 89–94.

Nordmann, A. 2014. "Responsible Innovation, the Art and Craft of Anticipation." *Journal of Responsible Innovation* 1 (1): 87–98.

Nordmann, A., and A. Rip. 2009. "Mind the Gap Revisited." *Nature Nanotechnology* 4: 273–274.

Roache, R. 2008. "Ethics, Speculation, and Values." *NanoEthics* 2 (3): 317–327.

Sand, M. 2018. "How the Future Has a Grip on Us." *Techné: Research in Philosophy and Technology* 22 (2): 262–282.

Sand, M. 2019. "On 'not having a future.'" *Futures* 107: 98–106.

Sand, M., and C. Schneider. 2017. "Visioneering Socio-Technical Innovations – a Missing Piece of the Puzzle." *NanoEthics* 11 (1): 19–29.

Selin, C. 2014. "On Not Forgetting Futures." *Journal of Responsible Innovation* 1 (1): 103–108.

van der Burg, S. 2014. "On the Hermeneutic Need for Future Anticipation." *Journal of Responsible Innovation* 1 (1): 99–102.

Weckert, J., and J. Moor 2007. "The Precautionary Principle in Nanotechnology." In *Nanoethics: The Ethical and Social Implications of Nanotechnology*, edited by F. Althoff, P. Lin, J. Moor, and J. Weckert, 133–145. Hoboken, NJ: Wiley-Interscience.

Wilsdon, J. 2014. "From Foresight to Hindsight: The Promise of History in Responsible Innovation." *Journal of Responsible Innovation* 1 (1): 109–112.

2 Hermeneutic Technology Assessment – Why It Is Needed and What It Might Be[1]

Alfred Nordmann and Armin Grunwald

Contemporary societies are obsessed with the contours of future. They do not ask what is wrong at present and how technologies can improve the situation. Instead, they witness technological change, expect more of it, and seek some kind of assurance that this will be a change for the better. Accordingly, TA (technology assessment) is called upon to develop scenarios or answer questions about the future of work or the future of mobility, about energy futures, industry 4.1 and the human being 2.0. This creates a theoretical, practical, and political predicament for TA which becomes apparent in the following statements, each with an air of paradox:

- The future does not exist as yet but is nevertheless treated as an object of planning and prediction, design and contemplation.
- Consequentialism is only one major strand of ethics (alongside virtue ethics and a deontological ethics of duty) and yet it is predominant in TA, even though the implications or consequences of emerging technologies are hypothetical at best.
- The art of anticipation involves a sensitivity for what can happen in the world as we know it, and yet we are asked to anticipate what can happen in a world very different from ours, one that has been radically transformed through technological innovation.
- Even though everyone understands that the future cannot be predicted, we hold on to softer versions like scenario construction, foresight, anticipation, imagination. And though we don't know whether this or that will actually be, many are confident that they know what plausibly might be.

Hermeneutic TA seeks to avoid this predicament and any attempt to evaluate emerging technologies by first imagining their consequences or implications. It considers the future as it appears in human conversations, popular culture, and policy visions, as it appears in calls for proposals and research applications, but also in prototypes and proofs of principle. Hermeneutic TA thus focuses on "the future" as it exists already – irrespective of the question whether there are identifiable paths from the

DOI: 10.4324/9781003322290-3

present to the future as it is envisioned. The futures in our present may have little to do, indeed, with any factual future state of humanity.

The conception of "hermeneutic TA" emerged from extended discussions of the promise and predicament of TA that seeks to provide some insight in what the future may hold. It was preceeded on the one hand by various proposals to avoid references to the future entirely in TA (Nordmann 2007, 2010, 2013a, 2014) and on the other hand methodological proposals to address possible future through an "explorative philosophy of emerging technologies" or by way of "vision assessment" (Grunwald 2010, 2014).

It remains an open question, however, whether and how "hermeneutic TA" can provide an answer to the difficulties of TA – or whether instead it adds to them: Can it do justice to the philosophical tradition of hermeneutics? In which way can hermeneutic TA still be technology assessment at all or will it end up, by necessity, providing at best a kind of discourse assessment? Hermeneutic TA is grounded in the recognition that our attempts to discover the contours of the future tell us only about the hopes and fears, practices and discourses of today. But does that mean that a "hermeneutics of the future" is limited to the meaning of utterances or texts – which would leave it indistinguishable from discourse analysis and interpretive culture studies? Do these explorations merely serve to turn the gaze inward such that the original attempt to look ahead into the future ends up a kind of methodological navel-gazing? Can there be something like "hermeneutic TA" that ends up assessing, perhaps critiquing *technological* projects, projections, and proposals?

In order to address these questions and to articulate a substantial notion of hermeneutic TA, Armin Grunwald and Alfred Nordmann assembled a group of scholars that first met in Karlsruhe in January 2016.[2] From these discussions it seemed possible to extract or condense a strand of observations, comments, and suggestions. Though it most certainly does not provide a "consensus" of all who were present, it affords a conception of "hermeneutic TA" that takes seriously each of the three terms: a *hermeneutic* method to *critically assess* proposed technological projects.

1 The archaeological work of present-day TA unearthes as yet unrealized technological project(ion)s and technological systems, including artefact and social designs, prototypes, and the products of visioneering.[3] Like the archaeological objects that come from the past, these have a temporal order that sets them apart and renders them incommensurable with respect to the quotidian objects of the everyday. This temporal order can be captured by the notion of the "prospective ontology of technology" (Rip 2009) in the sense that these objects lay claim to a future way of relating people and things. This intentional structure of technological projects and systems is manifested in virtue of their existing now and simultaneously not-yet, and it can be

described in terms of their specific claims to extend human bodies or to make devices available, or to implicate humans in a technical system.

2 The prospective ontology of technology has been likened to that of a "hopeful monster" (ibid.) – Richard Goldschmidt's 1940 term that first described biological mutations and then came to signify technical innovations in the work of Mokyr and Rip. Whether we think of self-driving, autonomous vehicles or soft machines, grids for a wind- and solar-based energy system, in-vitro meat or ambient intelligence devices, these hopeful monsters are a product of their time and have their time inscribed in them, but – like artworks or archaeological artefacts – they cannot be seamlessly resolved into the background, they are continuous and discontinuous at once, they do not achieve unity or totality but expose tensions, dreams, desires, hopes, fears, conflict, and contradiction. As such they are inexhaustible and require an effort of listening and reading, that is, a hermeneutic analysis that opens them up – in contrast to an interpretation that closes them down. Such an analysis will exhibit these hopeful monsters as *loci* or sites for the production of futures in the sense that each configures present capacities and each suggests new ways of relating people and things.

3 This listening and reading, probing and questioning, this opening up of futures takes place in a specific context. Systems of law, for example, institutionalize certain habits of mind, patterns of expectation or routines of judgment. Accordingly, these systems are a sensorium of sorts. Having internalized them, we have a sense of how flexible they are and thus for the difficulties that lie ahead when it comes to accommodating new ideas and things within these institutionalized ways of thinking. It is these institutionalized or frozen habits of mind that "anticipate" novel objects, systems, or devices, that is, they embody the sociotechnical imaginaries (Sheila Jasanoff) which define or circumscribe the world in which these promising objects might grow up (Jasanoff and Kim 2015). Sociotechnical imaginaries frame the questions through which we query and hermeneutically understand and seek to "anticipate" the objects that are poised between past, present, and the promise of a future. What matters here is not that one can actually anticipate but that one trains the understanding by attending to the *loci* or sites in which sociotechnical imaginaries operate and promises of the future are produced.

4 The discourse about emerging technologies and "our (technological) future" is entangled with the sociotechnical imaginaries, that is, with discursively and institutionally sedimented habits of mind which need to be questioned regarding their particular ways of visioneering. This is true not only of the objects but also of the subject of hermeneutic analysis – who are the "readers" of technological programs or prototypes, how are they constituted and provoked to change? The

opening up of texts and envisioned devices goes along with questioning the role and identity of the readers or users. Hermeneutics as a methodological practice mobilizes the critical subject and producer of meaning against the implicit "we" of institutional and symbolic orders. This opens up space and produces a critical self-understanding of human entanglement in sociotechnological systems – and it thus contributes to hermeneutic visioneering assessment.

One example of hermeneutic TA might concern an assessment of nanotechnology's so-called "soft machines" (Nordmann, this volume chapter 11). Nanotechnology promises disruptive, perhaps radically new technical possibilities. Accordingly, the machines it envisions are not only incredibly small but designed in entirely new ways. The dream of nonmechanical machines that are perfectly integrated with their physical environments proves to be a dream of a whole new working order for people and things. In all likelihood, the soft machines of nanotechnology are technically impossible and will never be realized. In the meantime, their hermeneutic analysis reveals human ambitions for a transcendent future order of things.

Notes

1 This paper was first published as Nordmann, A., and A. Grunwald. 2017. "Hermeneutic Technology Assessment: Why It Is Needed and What It Might Be," in Contours of the Future: Technology and Innovation in Cultural Context, edited by D.I. Kuznetsova, V.V. Sergeeva, N.I. Almazova, N.V. Nikiforova, 21-25. Saint-Petersburg: Asterion.

2 Aside from Grunwald and Nordmann, Arie Rip and Martin Sand, the group included at its first meeting Vincent Blok, Marianne Boenink, Christopher Coenen, Sascha Dickel, Arianna Ferrari, Steve Fuller, Ben Hurlbut, George Khushf, Sebastian Pfotenhauer, and Christina Schües. Though she could not attend the meeting, Sheila Jasanoff was very much present in spirit.

3 This could be a definition of "visioneering" which expands upon Patrick McCray's original formulation: Visioneering mobilizes skills, expertise, and resources to forge something much stronger than a narrative thread that more or less plausibly leads from the present to the future. What visioneering aims for is to exhibit a compelling causal link between a state A (technological work-in-progress) and a state B (a future so desirable as to mandate its realization) such that A will actually lead to B while B necessitates A. This definition calls for a more rigorous formulation in terms of Aristotle's conception of causality: Visioneering establishes A as the effective cause of B and B as the final cause of A. It thereby first seeks to identify an actual causal chain of events leading from A to B, and then offers B as a teleological end that is productive of this causal pathway as its means. Now, the notion of "Visioneering Assessment" would designate an attempt to probe the quality of these constructions – as such a further development or specification of "vision assessment" as conceived by Armin Grunwald (Nordmann 2013b; cp. McCray 2013).

References

Grunwald, A. 2010. "From Speculative Nanoethics to Explorative Philosophy of Nanotechnology." *NanoEthics* 4: 91–101.

Grunwald, A. 2014. "The Hermeneutic Side of Responsible Research and Innovation." *Journal of Responsible Innovation* 1: 274–291.

Jasanoff, S., and S.-H. Kim, eds. 2015. *Dreamscapes of Modernity: Sociotechnical Imaginaries and the Fabrication of Power*. London and Chicago: University of Chicago Press.

McCray, P. 2013. *The Visioneers: How a Group of Elite Scientists Pursued Space Colonies, Nanotechnologies, and a Limitless Future*. Princeton: Princeton University Press.

Nordmann, A. 2007. "If and Then: A Critique of Speculative Nanoethics." *NanoEthics* 1: 31–46.

Nordmann, A. 2010. "A Forensics of Wishing: Technology Assessment in the Age of Technoscience." *Poiesis & Praxis* 7: 5–15.

Nordmann, A. 2013a. "(Im)Plausibility Squared." *International Journal of Foresight and Innovation Policy* 9: 125–132.

Nordmann, A. 2013b. "Visioneering Assessment: On the Construction of Tunnel Visions for Technovisionary Research and Policy." *Science, Technology & Innovation Studies* 9 (2): 89–94.

Nordmann, A. 2014. "Responsible Innovation, the Art and Craft of Anticipation." *Journal of Responsible Innovation* 1: 87–98.

Rip, A. 2009. "Technology as Prospective Ontology." *Synthese* 168: 405–422.

3 Future Conversations – A Topical Exchange

Vera Borrmann, Christopher Coenen, Alfred Nordmann, Roberto Cantoni, Sascha Dickel, Christopher Groves, Armin Grunwald, Ben Hurlbut, Sheila Jasanoff, George Khushf, Sebastian Pfotenhauer, Arie Rip, Martin Sand, and Christina Schües[1]

On the Time, Times, and Temporality of Things

Alfred Nordmann and Armin Grunwald:

> The prospective ontology of technology has been likened to that of a "hopeful monster" – Richard Goldschmidt's 1940 term that first described biological mutations and then came to signify technical innovations in the work of Mokyr and Rip. Whether we think of self-driving, autonomous vehicles or soft machines, grids for a wind- and solar-based energy system, in-vitro meat or ambient intelligence devices, these hopeful monsters are a product of their time and have their time inscribed in them, but – like artworks or archaeological artefacts – they cannot be seamlessly resolved into the background, they are continuous and discontinuous at once, they do not achieve unity or totality but expose tensions, dreams, desires, hopes, fears, conflict, and contradiction. As such they are inexhaustible and require an effort of listening and reading, that is, a hermeneutic analysis that opens them up – in contrast to an interpretation that closes them down. Such an analysis will exhibit these hopeful monsters as loci or sites for the production of futures in the sense that each configures present capacities and each suggests new ways of relating people and things. (Nordmann and Grunwald 2017, this volume chapter 2, citing Rip 2009)

SCHÜES: "These hopeful monsters have their time inscribed in them" – this is true for living beings, in-vitro meat or bacteria. Does it make a difference whether this is in such beings or in an artwork, a text or perhaps a theatre play? Would it make a difference if these were also affected by time? A stone is not affected by time in this sense, it is not sensible to time, although the meaning of a stone may change over time.

DOI: 10.4324/9781003322290-4

NORDMANN: "Time inscribed" refers to the makers who produced the prototype and inscribed into it a future world. For these artefacts it is their nature to be carriers of the future. Part of the nature of these artefacts is to be sentinels or harbingers of the future.

SCHÜES: Bacteria might develop a life of their own in the lab. This kind of temporality built into the prototype itself will prevent that the maker can still claim: "This is what we made!" Do we have to take some kind of materialistic turn when we speak of temporality and time?

GROVES: This distinction is very important – as is, probably, the concept of "timescape" here. We have expectations concerning objects. Technologies (e.g. nuclear power) have their own timescapes. On the one hand, there is always a future inscribed in technology. On the other hand, there is also always an inescapable potential in technology as it unfolds in time, e.g. concerning a self-sustaining economy.

SAND: *responding to Schües, her emphasis on living and non-living beings:* I try to understand the difference between inscription into living and non-living beings by means of reference to "Frankenstein." Does the being have two time inscriptions – as a creature and living being, and as a being brought to life by Frankenstein?

SCHÜES: A being has a self-perception of its own time. Bacteria have no self-consciousness, I would think; but we inscribe temporality. Do we need a materialistic turn concerning beings that are not conscious of themselves? Temporality is inscribed by us, the perceivers. If Gadamer is right, only humans have a sense of the future.

KHUSHF: These questions of temporality are very interesting. My question is: Let's say we take a cellular system and we have time inscribed there. We take it higher, first organism, then multi-cellular, then move it still higher to something like a community (multiple individuals and more complex beings). Is the nature of this inscribed temporality altered by the scale? Do the concepts allow for these moves in terms of scale? Is the distinction living/non-living really helpful here?

SCHÜES: I rather hesitate to talk in the way of vertical scales. When Scheler and Husserl talk of more and more complex beings that, when considered at the highest level, have a mind; all is nicely built up in a kind of a vertical, even hierarchical structure. I don't think we need this kind of traditional thinking. The interesting question concerns temporal practice vis-a-vis the individual and relations. If it comes to "do we have to distinguish living and non-living?" we have to keep in mind that the talk of communities of living things also having a life is metaphorical.

KHUSHF: This is not yet an argument made by me, but at the moment only a play of thought: If you have the layered account, you enter at many levels. Structures of temporality at higher levels move down and vice versa, bottom-up or top-down structures of temporality are broader, compare perhaps Gadamer's account of a fusion of horizons because it presupposes that experiences, actions, or practices have in the end all compatible styles and forms. Difference (Derrida: *differance*) would get lost.

SCHÜES: You always need more than one type of criterion – the distinction living/non-living is not one of complexity. I would be very cautious to talk about a fusion of horizons.

HURLBUT: With Heidegger one might ask "Does a stone have a world?" – *Dasein* as world. Stones are world for us, and someone may one day say: Actually, a stone has a world. In the short paper, immediately after the passage we discuss now, Nordmann and Grunwald write:

> However, this listening and reading, probing and questioning takes place in a specific context. Habits of mind are institutionalized in systems of law, for example, and these systems are a sensorium of sorts for the difficulties of accommodating them. It is these institutionalized or frozen habits of mind that "anticipate" novel objects, systems, or devices, that is, they embody the socio-technical imaginaries (Jasanoff and Kim 2015) that circumscribe the world in which these promising objects might grow up.
>
> (Nordmann and Grunwald, this volume chapter 2)

If we think about the "hermeneutic" in "hermeneutic TA (Technology Assessment)," it is important to keep in mind that making sense in a phenomenological way is very complex and situated in history. I very much like the notion of sensorium used by Nordmann and Grunwald in this passage. This is not *Dasein*, ahistorical, but much more situated, historical, complex, and contingent. And I have a question: Is technology already social for everyone in this room? Hans Jonas pointed to a shift from the understanding of the living to the notion that we can create everything bottom-up from dead material in the construction of life.

PFOTENHAUER: The general paper about hermeneutic TA and Grunwald's input "What is the current practice of TA?" gives rise to several questions. First, the short paper nicely brings our discussions together, but if we take social construction seriously, then we will have different approaches to hermeneutic TA within technology assessment, viewing the future as socially constructed with a view also to the way in which we construct technology assessment. Missing in this, secondly, is perhaps a sense of playfulness, tinkering with the future, tentativeness. Then, thirdly, what is the archetypical problem of TA? Is there a situation that requires or makes the case that hermeneutic TA is needed? Is there a focal point, a case in a nutshell? And finally, what about epistemic access to the future and the political effects of TA as a social and political assessment? If we flip hermeneutics upside down, world-making and a reflection of the social order will lead to a "hermeneutic social future assessment."

RIP: I am impressed by Sebastian Pfotenhauer's laundry list, the dirty things. What do we mean by hermeneutic TA? I did an exercise in a chapter a while ago (Rip 2012): There is TA methodology, then a kind of ontology, and then something like a philosophy of technology assessment which is reducing the damage of technology to humanity by anticipatory learning and feeding back the learning in action and decision making. "Technology Assessment" is an umbrella-term with philosophy in one column of the table and in another column methodologies – a strategic alliance in the risk society. Is there anywhere in here a special place for hermeneutic TA?

COENEN: TA is frequently seen, also here at the ITAS Institute for Technology Assessment, if not as a discipline, at least as a field of research – but TA can also be deemed a general human activity based on the fundamental character of the human-technology relationship. I have started to play with this latter notion of TA, also because others, often from outside academia, explicitly engage in "technology assessment" now.

No Ethics without a Future

George Khushf:

1 "Hermeneutic TA" appears to be just one kind of TA, a form of TA focused on its "soft," "humanities-like" side. However, as I understand TA, it is interpretive to its roots. As a result, I find the term "hermeneutic" to be especially apt as a characterization of a reflexive activity (explicitly thematized) in relation to the interpretive activity of all TA, a pervasive interpretive activity that is often implicit and hidden in the "harder" social science like language that wants to presume mere description. On my reading, *"hermeneutic TA" ought to be a general heading for all forms of TA* insofar as they are reflexive and thus put into play engagement with their own interpretive activity as part of their analysis of technology.

2 The *assumptions about the nonexistence of the future* (e.g. in nanotechnology discourse) miss one of the most crucial aspects of assessments of technology that ought to be explored by means of hermeneutic TA: namely, the *nature of time, temporality, future/present/past*, and the ways these both inform and are constituted by means of our interpretive practices. This would be a dismissal of the whole alternative tradition concerning these topics (closely associated with the humanities and hermeneutics) as illusion-oriented and illusion-based. Hermeneutic TA (properly framed)

might itself become the locus for working past the contrasts between philosophical and scientific accounts of time.

3 One of the aspects of hermeneutic TA I find especially problematic (in its agenda framing) but especially promising under my more general framing concerns the ways hermeneutics might open again a space (literally and figuratively) between nature/science and politics/technology. The aim would be to uncover the ways that *all activities of science are "ethical" to the core*, and all visionary programs of technology-formation (like those that are the target of hermeneutic TA for Nordmann and Grunwald) are not just "political," but also scientific and oriented toward a new regime of facts and science as well as based upon, thus justified, in relation to the old regimes that bootstrapped us into our new regimes.

When we discern the ethical character of all scientific activity, we can speak about "*responsible science*" in a genuine way. *Without a future, there is no ethics and no ethical analysis.* In that case, all that is "political" degenerates into an ad hoc clash of power and justifications that reflect nothing but these configurations of power and their legislation. By ethical analysis, I mean an analysis that oriented toward altering what can be otherwise, so choices about such alteration can be made in positive rather than negative ways. Putting my worries in a more direct way: Are we assuming we are purely and accurately descriptive when we analyze agendas of emergent technology like nano and state that they are just (or primarily) "political," that the futures they present do not exist, and these are contrasted with mature science and technology where a more epistemically rigorous kind of noninterpretive TA is possible? Or are we more honest if we *acknowledge that this frame of ours – associated with the agenda of hermeneutic TA presented – is itself political* or at least reflects our take on the science/political divide, allocating initiatives like those of nanoscience to the "political" side? Here I will be frank in wanting to *open up a middle space between science and politics* where there is a recognized liberty and middle space for formation of multiple communities with different visions. I want to defend ways scientists (and "technologists") gather to advance things (indefinitely thematized in terms of general social, democratic agendas) they find interesting and worth pursuing, and the diverse ways they seek to mobilize support for their agendas. How these activities gear into the past is something worth pursuing as part of a hermeneutic TA. To this extent, *I argue for a variant of what might be called "classical liberalism,"* although I would reject the ways this is associated with a specific kind of capitalism itself interpreted in terms of a purely consumerist logic in which the diverse ends, epistemic practices, and ethical forms of life I would advance cannot be defended. However, in arguing for a middle space, I want to clarify a notion of a responsible practice of communities that emerge in these middle spaces and link an absence of responsibility in their self-governance (the free practices) to that which justifies mobilization of political processes which are (on my rendering) now circumscribed and more closely linked to the logics of force, constraint, and redirection that are often taken for granted and presumed legitimate over too great a range. *Hermeneutic TA can be a privileged, natural locus for re-opening and exploring a rich set of questions about the ways future, freedom, and novelty are intertwined and arise in relation to the interpretive activities* that both depend upon and rework the ways we experience time and temporality. (Khushf 2018, unpublished)

NORDMANN: What is the alternative version that is implied in Khushf's critique? We have two dimensions of our problems. First there is the question of how to access or relate to the future. Here, the premise is that the future is "*unverfügbar*" – it is not available to us as an object that we command or control. We can ask if we can actually know the future against the background of Khushf's points concerning epistemology and ethics: What can we predict? We can also note that actions are always oriented toward the future, they take place in the horizon of the future, but all the while, the future remains *unverfügbar*.

And then there is Khushf's second critical point – that we do not appear to believe in the existence or reality of the future. Here, the premise is that the future is a mandate or aspiration but becomes trivial if it is taken as an existing reality. This calls up considerations of history and time, questions about world-making and contingency: We lose the openness of the future if we take it as an object of design. This "non-existence of the future" is best understood if we say that the paper by Grunwald and Nordmann is based on the suspicion that the notion of the future has been rendered trivial. Taken seriously, future-making should be understood as world-making by a historical subject under the condition of contingency: We are always, in fact, making the world but we are never able to design the future.

RIP: We could ask about hermeneutics, NEST and hermeneutic TA: Is hermeneutics TA? What does this have to do with NEST – with New and Emerging Science and Technologies that claim to break with the past? How does the area of NEST relate to something in the past?

KHUSHF: I think we agree that there is something novel here that resists established modes of TA. It is important that our critique of NEST discourse uses hermeneutics as a means to deal with continuity and discontinuity, and also with applied science and technology. But this does not call into question the basic point that technology assessment always involves hermeneutics.

With respect to the remark that "the future does not exist as yet but is nevertheless treated as an object of planning and prediction, design and contemplation," an interesting question is *how would we think of the future as already existing.*

HURLBUT: Khushf presents a very rich set of questions, also concerning the configuration of what Nordmann and Grunwald called "sensorium." The problem is, however, not whether the future exists or not. NEST have always time built into them, if only as objects of concern that demand responsibility for the prevention of catastrophes. And though the notion of responsibility always calls up an underlying metaphysics, isn't it liberating in hermeneutics that you don't need a "once and for all"? No need to worry about purifying terms? We got 30 or 40 years of science studies telling us how important the context of facts is, etc. Can we take this for granted, also regarding the question concerning the existence of the future?

PFOTENHAUER: As regards implicit metaphysics, we can ask: When we are claiming that the future is inexistent, do we have different conceptions of the future? As a self-designated radical constructivist relying on 30 years of STS, I find Khushf's questions interesting, although I am in the opposite camp, not focusing on epistemology, ontology, or metaphysics but on politics and institutions. We can ask: Is hermeneutic TA referring to a technical practice? Is it really necessary? Does it help us to talk about what we want to do? And is it necessary at all to refer here to TA?

SCHÜES: We have here an epistemic concern related to laws, systems, institutionalizations, and so on. And an ethical concern as pointed out by Khushf, while the term "ethics" is apparently missing in the short paper by Nordmann and Grunwald. Khushf argues that if we do not have a future, we do not have ethics. It seems that this assumption is based on a focus on individual activity not of a process. But science can be deemed a process which sometimes does not provide a choice, one action following the other. Responsibility should be involved, but how? In the short paper by Nordmann and Grunwald, individual action is largely missing. It seems as if the paper by Nordmann and Grunwald and Khushf's comments play in two different dimensions.

KHUSHF: I wouldn't see myself seeing ethics in an individual way. Take as an example the diagnosis of cancer, where all kinds of factors play into individual decisions during the process. I also have a problem of putting everything on the level of the political. My question is how we can open space for middle-level questions (ethical, epistemological), between aspects of science hardwired and the political space (power). Without the middle level, at least I cannot make sense of the discourse on politics in the fully constructivist mode. One always starts in the middle, then you can move upward and downward. I take the relevance of all these TA topics for granted, like everyone here, but I think the integrity of the middle ground is important. There is always the presumption of descriptive accuracy in the social sciences, one always presupposes freedom, ethics, and so on. But on the mere presumption, we risk losing both, freedom and politics.

Mundane Futures

Christopher Groves:

The hold influential future visions can exert over the imagination depends on plausibility and desirability of futures. These are not just based on the cognitive coherence of information about possible futures, as a conviction that a particular future might be plausible or desirable can also be an aesthetic effect of how narratives about the future are put together (Eidinow and Ramirez 2016). People may question future visions based on a feeling of "wrongness" or alienation as much as on, say, questions about the provenance of the information on which they are based. Creating such aesthetic effects can then open a space within which the seams of future narratives can be teased apart. (Groves 2018, unpublished)

CANTONI: Christopher Groves engages with the ways in which people's hopes and desires for the future get bound up with past achievements. These past achievements are represented by technological systems which also point ahead to promising futures. But people may question future visions also concerning a sense of what feels right or wrong. Reflexivity about future visions can also be uncovered within everyday practice, emotional and ethical commitment and the narratives through which identity is produced.

Groves and his collaborators exemplified this in their "Energy Biographies" project, developing methods for exploring how the lived future enters into relationships with technology in the present, helping to create aesthetic and cognitive reflexivity toward energy futures. These lived futures, these mundane engagements with futures, can be employed in setting up an agenda for hermeneutic TA. Technology assessment needs to move away from considering "future developments of technology" toward reflecting on "future societal developments with technology."

But how one can operationalize this suggestion? This would be my first question to Groves. It is closely associated with my second question. Every time I have participated in a conference about applied anthropology or sociology the complaint I've most often heard was about the difficulties of social scientists to really have a voice in TA. How can this be changed? More generally: Since aesthetic ways of relating to the future cannot be measured, how can we speak of two groups of people with different aesthetic sensibilities coming to a compromise?

My final questions to Groves are: In which way are these imaginaries nation specific? What collectives are we talking about, and who holds those visions—sociotechnical managers, technological selectors, politicians, citizens? Is it legitimate to argue that all these imaginaries coincide? Thanks to scenario building, possible futures are multiplicated and the landscape of expectation is enlarged – but only in very specific directions, namely those envisioned by actors who have the means to produce scenarios?

GROVES: Reflexivity is key to technology assessment and the concept of hermeneutic TA, also the empirical research that comes with the hermeneutic approach. What can technology assessment be? Here the work of Mike Michael is important as we ask how do people enter TA? How is identity bound up with people and their sense of the future and technologies? Here we always find that the uninvited public is already engaged anyway. And when we expect people to engage as citizens, we forget about their emotional baggage:

> […] reflexivity towards socio-technical change is a property of concern which is also informal and uninvited, as much research in science and technology studies (STS) has shown. Rather than just arising among strangers within the formalised spaces created by engagement practitioners, it can arise out of processes of collective identity formation, as common commitments and care motivate the formation of collectives. Indeed, such commitments and identities that are dependent upon them are often the means through which engagement with socio-technical change occurs in the first place, rather than being an obstacle to it. More broadly, engaging with socio-technical change is something which can be seen as woven into the warp and weft of everyday life. (Christopher Groves 2018, unpublished)

We want to mobilize aesthetic engagement because we are interested in how people justify particular practices and how people are organized toward the future. Aesthetic aspects are also important for the ways in which people think ethically. There is an issue about balancing values. What might a good life look like? The methods we have been using are about opening an ethical, a moral space. Whether consensus is possible through this is a completely different question but some of the issues we have talked about might also touch upon questions where people might be ashamed (their use or waste of energy, for example).

RIP: It would seem that your work will benefit from narrative analyses practiced in psychology.

HURLBUT: Also, of course, from socio-technical imaginaries. What one could ask here is: "What is the perception of what a good life is?" But market research also asks: "Which technological world would people like?" I think lifeworlds and futures should not be reduced to preferences of different actors.

GROVES: With respect to market research, there too we are engaged with aspirations to make the world better. A typical narrative is the one about "smartification." There will be a smart world, everything will be better. This has become something like a trope that is not owned by anyone and that is configured differently by different actors with different interests.

NORDMANN: Christopher Groves also calls up the mundane questions of anxiety and hope as they appear to the uninvited, asking how they orient us to technology, serving as a sensorium in a future-directed way. George Khushf argued earlier that if there is no future, there is no ethics. What are we to make George's point in respect to Christopher's concern with the "mundane" or "worldy" and its relations to the future? As for the mundane, we can distinguish between the world we actually inhabit now and the imagined world of the future, and then there is the question of how these worlds get related by way of anxiety and hope? For example, to what extent is the anxiety that the world might end in a nuclear war a vision of the future? Isn't it much rather part of our mundane, present-day existence? If I think about the nuclear age, what is now at stake is the further existence of the present-day world – the future is nothing but the "not yet" of the actual manifestation of a clear and present danger. So, is our anxiety really oriented towards this future occurrence of our current capabilities?

Having a Future

SAND: There is an issue of power here: If we accept that there is a multiplicity of futures in the debate, how do we reflect this in the present? I want to argue that we have to understand the inequalities in future making. We asked already what kind of knowledge is generated from the perspective of hermeneutic TA. Can this knowledge inform policy- or decision-making? Looking at public discourses we encounter present futures which are often "vanguard" visions. Many are premised on the assumption that even in the face of open questions and risks, people will choose to adopt new technologies (Carlson 2010, 44). According to Carlson, no fundamental barrier to this technology is set by physics, chemistry, or biology – the pace of progress will depend only on the rate at which we reduce our ignorance about how to build devices. By the same token, we encounter a lack of "mundane" visions – in contrast to Groves's conception I mean by this the visions of ordinary or socially excluded people. According to Jasanoff, "[...] most of the world's masses are in no position to anticipate for themselves either immediate benefits or improved long-term prospects from the forward march of technology. They must accept the promise of benevolent outsiders that their lives will be bettered through inventions designed elsewhere, by entrepreneurs closer to technology's moving frontiers, with the capital and know-how to engineer large-scale change. Inequality – not only as access but even more of anticipation – thus emerges as an unresolved ethical and political barrier to the just governance of technological innovation" (Jasanoff 2016, 255–256).

The following questions may arise for our further discussion: (a) What does "having and not having a future" mean? (b) These mundane, disenfranchised visions exist, but they never come to the fore of public discourses – is this a "blind spot" of technology assessment? (c) Do these visions remain unnoticed because they lack a technological focus and are therefore neglected by the dynamics of capitalist societies? (d) How is optimism or hope regarding the future and the willingness to entertain visions affected by social standing? (e) If we agree with the diagnosis of an inequitable distribution of futures, how do we deal with this? By democratizing future making? Should anticipatory capabilities be promoted? (f) How does this relate to hermeneutic technology assessment? Does it simply add to "critical discourse analysis," especially suitable for discourse about the future?

RIP: Again narrative psychology provides some answers. Our problem is much broader than the "technological focus." The question could therefore be changed to "What does it mean not to have a future?"

SCHÜES: What does it mean to lose the future – this question applies to old people as well. One's own life and conduct do not fit anymore to visions of "future," and yet old people might still have a future. How does this question of not having a future relate to technology assessment? From an STS point of view there is no part of society that is not affected by technology.

PFOTENHAUER: Add to this the question where these visions come from. "New" types of visions are sometimes quite old, but there is a "violence of organized forgetting," that is, certain imaginations are systematically erased, for example by neoliberalism.

JASANOFF: Time is a very important factor here. Future is not only bound to economics. What do old people do, how do they react to the perspective that their lives will not last for too long anymore? When property and power become crucial for envisioning a future, this could lead to suicide in some cases. But in religion, for example, there is the hope for reincarnation. The catastrophe to die or to be poor is not the end. It is important to consider how people value time. The choice of temporality involves questions of justice. The science fiction genre plays with that. What do you choose to give prominence to? A good example is the election of Donald Trump in 2016. He stated "America will be great again!" There was a promise made to the "white left-behind men." Their wishes were taken to the future from the status quo.

HURLBUT: There is also talk about "the pace of progress will depend on …." One might say that some Enlightenment narratives about the formation or *Bildung* of the person as well as the capabilities approach view human development as involving the right to a future.

NORDMANN: When we are talking about different ways to value time, where does "having a future" come in? It would appear that to have a future is a possession among others – a possession that everyone ought to have, a marker of privilege, normatively charged. Tolstoy in his text on "Modern Science" turns this upside down: Science and its promoters rely on the future to deliver the promises of science and technology. They have a future but they lose the world, cannot answer the question of how to live today. Now, if we put these attitudes side by side – Lev Tolstoy and Martin Sand, if you will – shouldn't we say that, perhaps, that we share a world with "multiple temporalities."

JASANOFF: Maybe the future means also a "devolution" of technology. Maybe there are many different "planets" with different times.

NORDMANN: Sheila Jasanoff talks about a kind of violence in future-making that actually erases the imagination and vision of a better world. I agree with Martin Sand that it is extremely important to expose this power of the self-proclaimed future makers. If we then take on board Grove's experimental approach to the mundane ways of assessing existing and proposed technologies, we might safely say that the notion of a "shared world" is not a romantic idea: There is a fractured picture of the world, many points of view, but they make sense only in reference to the one world which is in the final analysis, our world.

JASANOFF: There are radical differences between the worlds that people are imagining. Take, for example, Bangalore as the "Silicon Valley" of India. Power is about saying: This is the right future and the best image for discourse, and if you don't agree, you are a dying species. What is a shared world? There are many individual wishes and desires for the future.

SCHÜES: The reference to old people was not about individual thinking of whether one has a future or not. To assume the perspective of old people is a critical tool for politics and society. "Old" in capitalist terms means being out of the world – further calling into question the presumption of a shared world. There is the hermeneutic idea of a fusion of horizons, but many people live with non-shared horizons. We always live in times, experiences and links that are temporarily structured and some practices lead to non-shared horizons. But we can criticize that: There is a world that can collapse because it is not shared. Whoever has the power about time has also the political power. It is not only about having a future but also about a shared horizon, and about not losing one's past.

COENEN: It is important not to give up on the notion of progress, or, if we do not want to use the term, to not give up on the hopes it stood for. A notion of progress does not need to be crude, as for example by seeing artisanship as old-fashioned in a future of fully automated industrial mass production. I find it remarkable that in this conversation no one as yet has referred to digitalization – a notion that everyone else is referring to these days. It is supposed to economically stabilize a global co-temporaneity of the disjunct (*Gleichzeitigkeit des Ungleichzeitigen*), making it transparent to more and more people, promoting a fusion of horizons. I sympathize with George Khushf's proposal to take up older traditions of hermeneutics. If technoscience is always already a hermeneutic activity, we may think about the devolution of technology and the other questions discussed here in terms of alternative notions of progress – if only because the old, intellectually outdated visions of progress still "reside" in the technologies which – like digitalization – are shaping the global economy and societies.

KHUSHF: There is no single future ("the future"), and I am critical about a notion of progress that emanates from a shared past. Consider this example from the ethics context: A Jewish orthodox women has amnesia – she forgot that she must not eat ham. She is brought to a hospital, everything is arranged with kosher food, but she eats the ham of her neighbor. She obviously has a future, but her past is gone.

SCHÜES: The notion of horizon has a manifold meaning: It refers, for instance, to the past that has been lived through but is lost. It can only be remembered in its historical transformation. Furthermore, horizon means something that is not yet spelled out, either because it is in the future, it cannot be grasped in language, or for other reasons. Many people live with non-shared horizons though we see it as one horizon. We should try to find a perspective that is not European or Western or US-centered in order to tackle the question of in- or exclusion. This notion of "horizon" allows us to avoid some of the problems associated in this context with the term "world."

PFOTENHAUER: I risk to be the heretic, but I want to come back to the question "How does this relate to the field of Technology Assessment?" We talked about inequalities, power and a better world, but I still wonder how hermeneutic TA is different from other approaches that also talk about power, a better world and inclusiveness? The challenging question remains: What is specific to hermeneutic TA?

SAND: It talks about power relations in future making and not simply about the future as a societal fact.

Prototyping

PFOTENHAUER: Sascha Dickel reminds us that prototypes are understudied – which might make for the beginning of a collaborative research project. Thinking about it, we may also think of Langdon Winner and the material politics of prototypes along with notions and practices of experimentation. One can ask: What vision goes into the prototype? But the prototype is not passive: Which future is actually accessed through a prototype? How are futures produced in prototyping? And what does this say about hermeneutic TA? Isn't hermeneutics limited to words and images?

DICKEL: We may need to expand our focus and the focus of hermeneutic Technology Assessment, but I would like to stick to two questions: How is the future expressed in artefacts? How are futures produced and how do they travel? Such questions are widely discussed in conceptual papers about TA but not in empirical studies. If we want to understand what the future means in our society, we need to study its means of production. Sebastian Pfotenhauer's comments point to another *desideratum* – understanding prototypes as epistemic artefacts or as boundary objects that produce shared understanding. It is easy for people with different backgrounds to come together and rally around these objects. Interviewees said that prototypes can speak for themselves. They do so by way of being something like containers: ideas go in and out. And not being passive means that a prototype does something with its designers in the design process. Hermeneutic TA is important here, one single future cannot be accessed. A prototype in the design field often presents a prototype of the future: It embodies a future and makes the future testable.

GRUNWALD: Prototypes could be a good field of study of the appearance but also of the disappearance of futures. How and why do prototypes and futures disappear again? We have prototypes in museums, for example, that represent disappeared futures.

DICKEL: Indeed there are many failed futures und prototypes.

GRUNWALD: I prefer to speak of "disappeared" rather than "failed" futures – the latter expression is too normative.

RIP: We earlier talked about the notion of a sensorium. The prototype is like a sense. It does not stand for a future, but it speaks about the future. It is a promise but also has requirements. In the design process it takes many iterations, many cycles until one finally arrives at a materialized artefact. Prototypes tell us why futures disappear again in this promise-requirements cycle.

DICKEL: Promises and requirements because prototypes come with a context of demonstration. You present the prototype to see what works or works not, to learn or to show off. I think we need to differentiate between different prototypes. There are those that say: look, here is the future! There are others which serve as a model or something similar for experimental purposes. They are, so to speak, harmless and do not have any power.

PFOTENHAUER: Is the prototype testing or demonstrating, perhaps positing the future? Is there a serious intention to learn what a better future might look like?

CANTONI: I have worked on fracking that was presented as a continuation of older technologies, but also as really innovative etc. The question is how it is presented.

DICKEL: There is a politics of prototyping of big projects like nuclear power also on a smaller scale. One can study how prototypes are presented to get crowdfunding. It would be fascinating to compare prototypes in big science and innovation with these small-scale prototypes.

KHUSHF: The question of power is relevant from a paper model to a fully fleshed functional prototype. Miguel Nicolelis' monkey with a computer-interface was a prototype of a working prosthesis, there is no doubt it works.[2] Does that mean that feasibility has been demonstrated? Is it now only a question of desirability? Or even a question only of: Why shouldn't this be?

SCHÜES: We have been discussing how hermeneutic TA addresses the social and the technological. What does it mean to say that a prototype actually works? Should we frame prediction in TA as a matter of true or false vs. prototyping as a matter of it works or it doesn't work?

DICKEL: An artificial setting is built in order to see and test whether a prototype works. This setting and the demonstration might become a part of the prototype itself.

GRUNWALD: Is there perhaps an ontological asymmetry between working prototypes and prototyping? The focus on "it works" may lead to a soft version of technological determinism. But we have to take into account all the social preconditions that feed into prototyping.

PFOTENHAUER: I disagree that the social counterpart does not exist here: social labs, living labs, etc. can be a kind of prototyping of new social arrangements.

DICKEL: This would have been my answer to Armin Grunwald: "*Reallabore*" etc. could be seen as means of social prototyping. What happens when you arrange and present it as a prototype? We encountered practices and self-understandings of practitioners along the lines of "Let's be political without bringing the whole political complexity in!" Ideas about prototyping are not homogenous at all. An influential and very interesting case is the meaning of a prototype in design activity. Here prototypes are often not isolated artefacts, but are something like a smart city. We do not think about what this city might look like in 50 years, the prototyping process is more like "Come together and build something!"

JASANOFF: This contrasting of materiality vs. discourse is an artefact of academic politics. All prototypes are social and material at the same time. Including the normative: "it works." The judgement that "it works" involves ideas of for whom, in what context. For example, does it work for the family – our entire social world is built around such prototyping where the family itself and other models are of social organization the prototypes. How can we avoid falling into traps here? If a family with two children in Berlin is a prototype, is the US constitution a prototype as well? As regards technological futures such as the futures of robots, a whole infrastructure is needed to evaluate whether a prototype works or not.

DICKEL: I agree with this, except on one point: Not everything can be built as a prototype. There are too many conceptions about what a family is or can be, for example. But I would like to examine how the arrangements of prototypes influence expectations. Let's not fall into the trap of one idea of technological progress!

JASANOFF: The word "prototype" itself is not applied only to things that look like a machine, ready for the museum.

DICKEL: We need a distinction at some time between a political and an actor's perspective. My focus is how this concept can be applied to things that are not technological.

JASANOFF: There are also people who were not considered as people by the US Constitution. We can consider them as human or non-human. This is a very material social prototyping.

The End of History and Politics

NORDMANN: Dickel spoke of the future as being communicated through prototypes. I suggest a slightly different wording, such as: "The future is discovered or unearthed in artefacts" – just as we unearth the past by finding artefacts, trying to reconstruct from them how people might have lived. Prototypes have a precarious ontological status. Just like isolated things from the past that were found here or there, they are here, right in front of us, but they testify to a way of living that is quite alien to us, unfamiliar. The prototype is a foreign body in a very concrete, tangible way. As such it can question, undermine our current ways of living. At my university we have engineers who develop prototypes for the collection, purification, recycling, drainage, and disposal of water. But, sadly to them, our cities are hundreds, sometimes thousands of years old, they are a built infrastructure for the flow of water through the city, bolstered by trades, laws, regulations. How will the novel prototypes ever "work" in these old structures? So, in fact, we are confronted with a historical process as we work our way through from here to there. And then there are the prototypes that originate from an unknown place. With Ernst Bloch, perhaps, the prototype can be deemed a cipher of the future, the *Vorschein* (foreshadowing) of another, perhaps better world. I recommend Frederic Jamesons's *Archaeologies of the Future* where he argues that Science Fiction (like other forms of scenario building) "does not seriously attempt to imagine the 'real' future of our social system. Rather, its multiple mock futures serve the quite different function of transforming our own present into the determinate past of something yet to come" (Jameson 2007, 288).[3] As Richard Saage has argued, technoscientific visions aren't utopian but take the form of utopian thinking – only the form as they posit what the future will be like.

DICKEL: I like Jameson's book, I read it for my PhD on utopianism and human enhancement. There is a vanishing of utopia as a political idea for living and being together. This may have something to do with the acceleration analyzed by Hartmut Rosa which, according to him, leads to an instability of the present. We are unable to imagine in societal terms what another life should or would look like. Instead, we have the lure of technological progress. There seems to be an academic discourse proclaiming that there is no progress anymore, as mentioned earlier by Coenen. But actually, the notion of technological progress is very much alive.

COENEN: But skepticism toward this notion is strong in relevant parts of "Western" societies, for example the ecologically minded, "green" middle classes.

SCHÜES: How does it all this connect to a notion of hermeneutic TA? Does it require an awareness of contingency, and how important is the notion of the narrative? There is Hartmut Rosa's thesis regarding the instability of the present which states that narratives always need a location, need a present. Habermas's notion of discourse also requires a sense of presence and so, perhaps, does hermeneutic TA.

GRUNWALD: In my view, all TA is rooted in the immanence of the present. Take scenario-type thinking: Only a few parameters are changed, but in all other respects the future is like today. As concerns the "instability of the present," it is expressed by the totally different or radically discontinuous futures à la Nick Bostrom and the transhumanists. But such visions of the future or also, for example, Ernst Bloch's utopian views, do not make sense in TA and the world of scenarios. In a situation of deep instability, eschatological questions come back: Where does this all go?

SCHÜES: I agree, the Habermasian notion of discourse is perhaps more narrow than the narratives we are talking about, but both serve to stabilize the present as the place from which we talk about the future.

NORDMANN: We seem to agree that there used to be political utopias and now we have narratives about disruptive technologies, and these narratives are not rooted in a shared reality. Hermeneutic TA is a response to this development. I would like to come back to the materially present prototype as a herald of the future and how it comes up against a well-established infrastructure, as in the case of waste-water collection and its purification. There is a kind of inertia as a source of stability in all those ducts, collectors, and purification plants. These firmly entrenched infrastructures tell us: "This works, this is the way things are done around here." What TA can do is to revisit those simultaneously technical and narrative constructions, thereby showing how they constrain and qualify and render salient or make readable the future that is posited by the prototype as a foreign object of sorts.

DICKEL: In the modern world the future is always foreign. If someone would present a current city as prototype of the future city, we would find that completely irritating. As Reinhart Koselleck has shown, in modernity we always expect the future to be in some relevant ways different from the past. Maybe we should speak of "otherness" instead of "foreign object"?

PFOTENHAUER: I am not sure that I or, I think, Sheila Jasanoff would agree with this point about the city. Looking at the shiny glamorous cities of today, many will say that their children might lead precarious lives at the verge of poverty but why should the city of the future be different from today's? This prepares for my question to Armin Grunwald's distinction between TA and other, radically discontinuous visions of the future: If we take seriously the notion of hermeneutic technology assessment, we must adopt the STS insight that all kinds of change, big or small, are political, are significant, and that none are radically discontinuous. As concerns the question whether the future exists, we may ask: If it truly were radically different, how is it accessible?

HURLBUT: I am a bit surprised by the nostalgia expressed here for the lost capacity for politics and political utopias. The future always implies change with cultural significance. Referring to what Sascha Dickel was saying, Fukuyama's "end of history" came to my mind. No need for perfection anymore. The notion of "the future" orients us toward constructions of the future, it allows us to step away from the horizon of action that is available to present-day politics. But in the distancing from the horizon of action, there are our options. Even according to Fukuyama, his more recent book, there is a new action, a new movement toward a posthuman future.

NORDMANN: This is a very important point. And I agree, the notion of "the future" or, even better, the notion of the possibility of another, better world opens a space for the consideration of alternatives. In light of contingency, we can conceive of actions that might be productive of a better world – which does not imply and need not presuppose that we can know or shape the future. We can't even control or fully predict the consequences of our immediate intentional actions, and yet they are oriented to an idea of making things better. Notions of the "end of history" and the loss of the political concern not the difficulties of controlling or predicting what might happen, they concern this thinking and acting in respect to a better world. This loss of history and politics is quite compatible with, and might even be the by-product of thinking along the lines of trajectories only, of managing vulnerability, of solving problems technologically, including the problem of a nuclear threat or global warming. This is how Günther Anders spoke of the end of history – the nuclear apocalypse has happened already and we have no future anymore because we are living under Damocles sword of nuclear annihilation and from now on we can think no further than how to manage the risk and try to prevent the deployment of nuclear weapons (Anders 1980). I recommend the film "Never let me go" as a science fiction of sorts that is utterly stuck in an eternal present that we recognize as a past.[4]

JASANOFF: I wonder whether this narrative could work in Japanese culture in light of how it has changed and is changing.

COENEN: The nuclear apocalypse is not deemed the end of history, however, if the future is out there in space, an instance of a politics of technology as a kind of eschatology in action. As regards the idea of an "end of history," it is worthwhile to remember that Margaret Thatcher spoke of "There is no alternative (TINA)" already around 1980 when she was one of the first heads of government acting on the basis of what is now called neoliberalism. There is a loss of the open-ended future only in the current "neoliberal" capitalist world. The linear time of communist thinking with its promise of salvation is quite different from the idea of an "end of history," and so are Enlightenment notions of progress and perfectibility.

KHUSHF: Imagine I am a transhumanist, I am Elon Musk and have wild, radically different, not mundane vision of the future. But with my vision I am going to be confronted with all kinds of people with all kinds of other visions of the future. How we understand the dynamics of temporality at play in such confrontations is the most interesting thing in hermeneutic TA.

Time as Subject and Subject as Time

Christina Schües:

The concept of the future calls in question the history of ideas, and even various ways in which we go about our lives, ordering our behavior as a function of our conception of the future and of our relation to time. But how is this future to be understood? Is the future still open? Can it be derived from the past? Does it look threatening or friendly? Depending on what understanding of time we assume, the future appears in another light. Which leaves it questionable whether the future can even be grasped as a purely temporal phenomenon. [...]

Prognoses and Prevention as the starting point of technology assessment turn out to be examples of the consequentialist approach of our thinking about time. The relation to the future relies on prognoses and plans grounded in the present and is even extended to the representation of goals. As calculated, planned, and even imagined, the future is brought closer, and so takes shape as action in the present which is viewed across such feedback mechanisms as preventive measures.

But how can the relation to the future be grasped through the model of a multi-dimensional time? It refers to how time is experienced.

The future seems absent, and yet time is always there whenever we talk about it. This means we can't get behind or ahead of time in order to grasp it as a whole. This situation, Immanuel Kant calls the "a priori" of time and Paul Ricoeur the "inexhaustibility" of time. [...] Many phenomenological positions therefore remain in a framework of continuity, which, however, does not subscribe to a consequentialist thinking about time but to a conception of time belonging within the referential structure of what is co-intended.

As a third position, I would like to cover discontinuous conceptions of time. Following Emmanuel Levinas, concepts like "discontinuity," "novelty," "renewal," or "rebirth" or "restoration" stand for the dissolution of any philosophy of time founded in "finitude" (Levinas 1961 [TI], 281–285). That structure of time which forms the structure of consciousness rests upon an elementary gesture that "refuses" the totalization resulting from the collaboration of unity and being. This gesture is directed toward what cannot be unified, that, at best, has to be admitted as alteration, but that also shows itself as "Face" (*visage*, Antlitz) and, as such, conditions both time and consciousness. A consequence of this is that the "I" should be both "Engagement and Disengagement": It is disengagement from fantasies of what could have been and might have been accomplished, disengagement from regret of lost opportunities and the possibilities sacrificed by reality. And it means an engagement of the "I" with the drama of time, with a way of being that is spread between the manifold and discontinuity.

For Levinas and also Hannah Arendt, this calls for an attitude of forgiving. Arendt understands this "the possible redemption from the predicament of irreversibility – of being unable to undo what one has done though one did not, and could not, have known what he was doing" (Arendt 1989, 237). If the person whose action unwittingly hurt someone else, is forgiven, it puts that person in a position to make a new beginning. [...]

The constitution of a new beginning for the future world of human relationships is therefore based on a break with the continuity of the flux of time. Only this allows us to make sense of a beginning for our actions and communications. And only by requiring an explicit involvement with others, that is, by presupposing a plurality of selves and the impact of others on the self is it even possible to think the discontinuity of time and so to enfold the relation to the future as a "world-in-between" involving other human beings. Underlying the new beginning we therefore find a double condition: the break-through to the other as another and not as the same, and the continuation of a beginning by action with another – a beginning which will then emerge as a beginning because others have acted upon it. (Schües 2018, "The Relation to the Future and the Transcendence of Time," unpublished manuscript)

HURLBUT: Can we conceive of the future as a purely temporal phenomenon? A future can collapse in the face, for example, of a medical diagnosis and prognosis. Prognosticated futures demand action in the present. In medicine, diagnosis demands intervention, and diagnosis is future-oriented: Which future is appropriate? Relationally seen, a diagnosis is concerned with the being-toward-death, and yet in the case of a cancer diagnosis one asks: "How long do I have to live?" That is a question about ordinary time, a question about clock time. How do such consequentialist prognoses relate to a multidimensional understanding? Levinas and Arendt want to go beyond the subject. For Levinas, only the other and the relation to the other open a particular time. And in the meantime, we have the tendency to objectify, and from a hermeneutic TA perspective, maybe subjectivity is not even the problem at all, not an obstacle to be overcome.

SCHÜES: Concerning the question whether subjectivity is interesting for hermeneutic TA. What does "inter" mean in the concept of interrelation? For Heidegger and Merleau-Ponty it refers to "in between." If you take it one step further, the other has a claim on me. And here, I am already outside of myself, I am in a subject-subject relation. The technical subject is a subject as well, it is "embodied" – and time is embodied in it. Technology is not only an object any more. Care, love, forgiveness, etc. happen in between two subjects. Technology will have a subject position as well – that means an active position, not passive.

NORDMANN: Here is a question that opens things up even further. As we are speaking about phenomenology of technology and time, what about notions like patience and impatience, rhythm or beat, stagnation and acceleration, and other experiences with temporal structure? How much focus do we have on deeper technological structures in phenomenology and classical texts concerning time?

SCHÜES: In the work of Henri Bergson, we hear a lot about duration, the notion of infinity. These are notions which show a structure of the field and the in-between. In Derrida and Levinas, the emphasis is much more on how we are affected by the in-between. The structure, for example, of our transportation system. Hurlbut mentioned clock time. It gives our lives a certain rhythm. By watching the clock, we lost time already.

NORDMANN: I was also thinking of Ray Kurzweil's motto is "Live long enough to live forever" which one can also find in the discourse about climate change – hope to barely stretch things out so as to profit from future technological breakthroughs. There is something desperate about this, but also an eschatological form of religious expectation of deliverance, redemption, or salvation.

SCHÜES: In the condition of anxiety, I don't create something as a subject, but I am affected by it, receptive – and that is a shift. A novelty is novel only if somebody carries it forward. I can create something new, but what if nobody is affected by it?

GROVES: How do we think the intersubjective with respect to objects and their own temporality? In Günther Anders, the nuclear weapon exhibits the capacity objects have in them. The temporality of the object and the way it has a future in it seem very important for hermeneutic TA.

SCHÜES: This is about the temporality of the object and thus an object that is no longer a mere object – it produces not only a displacement of myself as a subject, it is also a de-objectivation. For certain assessments, however, we require a subject-object distinction. The challenge is to allow for a certain ambiguity.

GROVES: Indeed, not to flatten things.

PFOTENHAUER: With Levinas, responsibility is connected to the idea of being in the world as being for the other – infinite responsibility for the other. You cannot withdraw into your own subjectivity. This is a moral conception of the world. This brings ethics in a position prior to epistemology and ontology. Before we start posing questions of knowledge, relations, and interrelations are important. Everything is based on the relationships that already exist. If hermeneutic TA takes this into account before the assessment, it might flip everything dramatically.

JASANOFF: Interpretative frames are not "innocent." We do not have a choice but can observe the ironies, the ambivalence. This non-neutrality of the territory becomes apparent, for example, in Nordmann's remark on the point of living forever. It refers to a causal consequentialist point of view premised on an American resistance against ambiguity. We have to open up more to ambiguity and this takes us to the normative aim of our project, I think.

SCHÜES: Regardless or being oriented toward Europe or elsewhere, Levinasian hermeneutic TA asks: What world are we living in? The territory has an influence on which subject I will become. Time is within us, humans are sensible to time. And because of this, time can be understood as power as well. Because of this, humans can control other humans by having control over their time. Human beings are sensible to time and therefore can be disciplined through time, also the time of the future – all of which has a direct link to responsibility.

Anticipation-in-Action

Arie Rip:

Anticipation is now an integral, and institutionalized part of science and technology policy and practice – think of foresight exercises (including Delphi), roadmapping, and consultancy work (e.g. Gartner Inc.). But also promises are put forward in research and program proposals, and for newly emerging technologies generally (sometimes exaggerated). Philip J. Bond, the US Under-Secretary of Commerce, said in the Swiss Re workshop on nanotechnology in Dec. 2004: "Given nanotechnology's extraordinary economic and societal potential, it would be unethical, in my view, to attempt to halt scientific and technological progress in nanotechnology. The progress could enable us to feed the world's hunger, enable the blind to see and the deaf to hear. Given this fantastic potential, how can our attempt to harness nanotechnology's power at the earliest opportunity – to alleviate so many earthly ills – be anything other than ethical? Conversely, how can a choice to halt be anything other than unethical?" This can be unpacked factually – what are the chances that these benefits will be realized? – as well as morally: why the urgency, compare, for example, Hans Jonas "It is no sin to delay a benefit" regarding recombinant DNA research in the late 1970s.

This is not just a matter of actors anticipating and attempting to enroll other actors by promises. It might become a requirement on innovation to take existing roadmaps and other foresight results into account. Similarly, the repertoire of globally formulated and more specific promises is not something that innovators can just pick from. It creates soft requirements for directions to go, or at least how to justify the directions when they don't follow the repertoire (cf. the case of nanotechnologies). Of course, there are epistemological issues of prediction and its pitfalls. But there are also, and primarily, social and political issues regarding the embedding of anticipation – also long-term developments like the rise of modernity and its future-orientation. Therefore one might speak of anticipation-in-action. An example is the project FORMAKIN (2001). The project studied foresight exercises in the public and the private sector. It sees foresight exercises as an instance of "future-oriented coordination activities." Their role/impact can be pictured as an inverted U-curve. At one extreme, in a close-knit configuration, foresight is not needed because coordination occurs already, at the other extreme, in a loose-knit configuration, there is insufficient incentive to produce coordination as a collective good. Those in the middle profit from foresight exercises and develop anticipatory competencies such as Delphi exercises, invented in the 1960s by the RAND Corporation, popular for technoscience and technology since the 1990s. Here, experts and sometimes stakeholders in a domain are invited to offer by way of survey their views on the future of a domain. Their aggregated views are fed back to the participants to check possible reconsideration. The method is developed deliberately to determine whether a society has the capacities or competencies to achieve some goal.

The alternative entrance point is to turn the question around. Humans have a capacity for anticipation-in-action, some more than others. We can train this capacity, support it with tools, and delegate it. From early modernity, there are dedicated studies and specific forms of analysis, e.g. military strategy to anticipate your opponent, insurance deals to distribute risks in early-modern trade, the emergence of probability and statistics. These delegations go with packaging that allows anticipation to travel. Packaged anticipation includes historical data and extrapolation, probability calculations and estimates, informal and formal scenarios. There is also qualitative packaging in stories about the future such as early-warning stories. This also occurs "naturally" when people or organizations project possible and desirable futures, producing "actor worlds" in which expectation statements will be used as strategic resources (Callon et al. 1986). But they are not just strategic and their values lies in their content as well. Other actors, portrayed as characters in the story, will have to respond and they do so by offering a (partially) alternative story. The original author now becomes a character in the story of the new author, but the new author is not completely free in creating the story she would like, she has to take the original story into account. Such interactions continue, new actors may enter the arena. There will be some cumulation, not necessarily a consensus, but with some elements becoming more forceful than others.

There will be contestation and struggle in this kind of story-telling and thus a further dynamics. The struggle is also conducted by labelling opponents' anticipations as "speculative" and "science fiction." This reinforces the bias toward the cognitive, of anticipation being primarily a question of knowledge. There is also a professionalization of scenario-building in the 20th century that is originally linked with the military (cf. the RAND Corporation and scenarios about nuclear deterrence). By now, there is a small industry of forecasting and foresight, as well as roadmapping and technology assessment, serving private and public actors. Markets and audiences have to be reached and served with pressures to package, black-boxing the "art and craft of anticipation" (Nordmann 2014).

Now, my argument is not against delegation and partial professionalization of anticipation, including packaging, but about the lack of reflexivity in the action-orientation of anticipation. The effects of packaging – which is unavoidable – should be modulated by opening it up as a form of reflexive modernization, also by creating approaches that can accommodate complexity. This opening up of the narrative packaging will question that actors tend to project a linear future and project it from their intentions, using their projection as a road map – only to be corrected by "circumstances." They require forward-mapping tools that force actors to consider the non-linearity of co-evolution, and to accept the complexity. This can make actors more effective, if they are prepared to accept such social-science based support which can be done (cf. Robinson 2008).

So what about anticipatory coordination as a new stage in the dynamics of science, technology, and society? Perhaps we need a move to domesticate environments for future selection, that is, for the selection of futures that are unknown but deemed to become more manageable thanks to forward-looking coordination. This would involve reflexivity about the tools and approaches to do so. We need to unpack those and expose the basic mechanisms. (Rip 2018, unpublished notes)

SAND: You say that successful anticipation is not primarily an epistemic problem. But who has anticipatory capacity and who not?

RIP: The capacity to anticipate is a real question. There are different levels and dimensions of anticipations and also victims of disappointing dynamics.

GRUNWALD: Instead of asking about successful anticipation, we should ask, perhaps, what is a good anticipation? The criteria for good anticipation are manifold, and anticipation-in-action would mean: we should keep in mind for what purpose they are made.

CANTONI: I see especially one issue with the packaging of expectation and anticipation, namely that it tends to naturalize technological development. Technology is seen as an actor who drives something that will happen anyway.

RIP: We are a victim of our own perspective which underscores my call for reflexivity. Rather than naturalize an inevitable course of events, the Club of Rome's goal was to avoid that the results of their future-oriented analysis and their prognosis become true. And in regard to that they did a good job.

NORDMANN: Anticipation-in-action captures very nicely that we do not have to talk about anticipations of the future. But what is meant by anticipation when reflexivity is taken as a sort of anticipation? Isn't anticipation saturated with experience, isn't it our knowledge of the world that allows us to anticipate what might happen?

RIP: You are right, you can learn and improve anticipation. This happens also in the military as one learns how to anticipate the move of one's enemy.

KHUSHF: In preventive medicine, you want to look at some risk factor in such a way that you reduce disease. With what capacity can we make sense of anticipating, or is anticipation a capacity of its own?

The Role for Hermeneutics within the Practice of TA

Armin Grunwald:

I want to consider the assessment process, the conceptual core of TA, even what can be a definition of technology assessment in the end – and the place of "hermeneutic TA" or, rather, of an hermeneutic approach within this overall process. In order to identify commonalities among the many and various TA activities we need an overall perspective without losing contact with TA's foundation in practice. The crucial step in this respect is determining the epistemic interest of technology assessment. It allows the defining of technology assessment as a specific set of socio-epistemic practices serving this epistemic interest. From there, the objects of technology assessment will be determined as elements of its ontology, basic elements of its epistemology will be introduced, and the TA-specific steps of the assessment will be analyzed. The concept of epistemic interest was proposed by Habermas (1971). Generally speaking and across the various specific epistemic interests of TA, these would be its core elements: (i) recognition of the ambivalence of technology related to the emergence of non-intended effects, (ii) the opening up of new opportunities which should be uncovered as early as possible, (iii) a comprehensive systems view on the possible consequences of new technology, (iv) the organization of adequate processes of assessment, evaluation and shaping of technology and innovation with, (v) the expectation to open up critical perspectives on technology that complement the usually interest-driven views of the promoters of technology in politics and the economy.

As for the epistemic interests and how they changed in the last several decades, one has to take into account the increased influence of new science and technology on almost all areas of life today and in the future, attended by calls for more reflexivity and anticipation. Jonas's "Imperative of Responsibility" (1979) may be taken as the first grand ethical approach in this respect, followed by many others. By the same token, TA increasingly concerned with the assessment of possible, envisaged, designed, imagined technologies, with technologies that are not yet in existence and will perhaps never exist. The present manifestations of these technologies may be models, narratives, roadmaps, visions, scenarios, prototypes, demonstrators. Here, strategies of "trial and error" or "wait and see" are naïve with their assumption that new technologies are generally beneficial. By enhancing reflexivity to enable and empower society to better shape technological advance, TA provides a counter-model to these strategies.

Technology as such is not interesting to societal discussion and therefore also not for TA. Of interest are the challenges attributed to them. Since TA can often only assess "ideas" about future and embedded technologies, we have to extend the field to the techno-scientific nature of those ideas, thereby opening up spaces of possible applications because of their enabling character.

The traditional picture of technologies is that technologies are objects and humans the subjects, but TA should be aware of possible changes in subject/object relationships. It is the implications or consequences of technologies that are the objects to be assessed. The object of TA is the societal meaning assigned to technologies and socio-technical constellations possibly available in the future. This meaning is assigned in (at least) two forms: in terms of socio-technical futures and in terms of the characterization and definition of a new technology and its novelty or disruptive potential. The meaning is assigned by actors with specific diagnoses, perspectives, attitudes, values and interests in mind.

Here, finally, hermeneutics come in since these engagements and discussions implicate TA in a hermeneutic circle. This owes to the shift of epistemic interests as we reached the endpoint of the "upstream movement" in Science and Technology Studies, in applied ethics and TA which became prominent around the year 2000. The upstream movement comes to an end since the creation of socio-technical meanings of technology is the origin of the stream – this where we are, there is nowhere further upstream to go. There is only the circulation of cross-referential meanings. Hermeneutic approaches, but also media and literary studies or art history and anthropology, contribute to TA by analyzing this circulation of reference. (Grunwald 2018, unpublished notes)

NORDMANN: You said you had been reluctant to speak of "hermeneutic TA" – in the context of your remarks this makes sense. There may be quite a difference between hermeneutics as a humanities-dimension in a comprehensive form of technology assessment and conceiving TA as a whole in terms of hermeneutics – especially in respect to those changed epistemic interests.

GRUNWALD: I was nervous when I presented this because when we look back to the debate of the last 20-30 years, we see evidence that the old model worked. Introducing a broad notion of hermeneutic TA is therefore not a self-evident step. But now TA is to be increasingly applied to imagined technologies. We have to deal with actors who deliberately intervene in the hermeneutic circle. Hermeneutic TA is thus on a deeper level of reasoning than the old model and may increase our reflexivity. The role of TA concerning the hermeneutic circle is twofold: First, to analyze programs and prospects from a distance in a given time and context, and, second, to make a difference, to create an impact in the world. When the German parliament asked us to conduct a project on nanotechnology, the feedback created a new meaning of nanotechnology in our political context.

SCHÜES: You talk about co-constitution but sometimes also about application and consequences. I understand that you mistrust "application" as a term since it is not very reflexive. So, a hermeneutic turn might consist in this terminological shift to co-constitution, and what you say is more revolutionary than is suggested by the language of hermeneutic aspects or a hermeneutic side of things.

GRUNWALD: I have to think about this. Maybe "side" is too weak a term, but "turn" or "paradigm shift" too strong a word.

DICKEL: On your account hermeneutic TA does not stand alone, it is nothing onto itself. Hermeneutic TA raises attention to a hermeneutic dimension which was always there. With respect to the empirical side of hermeneutic TA in concert with other forms of analysis, we should talk about which disciplines provide what kind of understanding and how can they be brought together, without a privileging one's own discipline. In Khushf's view of hermeneutic TA, I like that it is about "understanding of different understandings." Actors in techno fields are also in the job of understanding – and we need to appreciate that in their jobs they are also in a reflexive understanding mode.

JASANOFF: As regards the idea of social meaning, one can notice that Germans are obsessed with reflexivity – a procedural notion, after all.

GRUNWALD: We associate these procedural concepts with Habermas and the tradition he speaks for. This extends to the emphasis on the inclusion of citizens which is thought to be a key element of reflexivity within TA.

HURLBUT: In discourse on inclusion and participation, they are usually seen as inherently "good." This is a problem perhaps not of the hermeneutic circle but of a hermeneutic spiral – reflexivity becoming an end in itself and spiraling out of control. In pre-hermeneutic TA, social meaning was seen as attached to the technology itself. And this was also not reflexivity, only a reflex, like when a doctor hammers on a knee.

GRUNWALD: As concerns the un-reflexive attachment of social meaning to technology, it is interesting to note that such attachments can drastically change: Human spaceflight was an important challenge with great symbolic meaning in the times of the Cold War. Nowadays it tends to be more about global cooperation, peace, etc., but this might be changing again as China and economic actors enter the game.

NORDMANN: It seems that Armin Grunwald as a TA-practitioner is wearing two hats: In the tradition of practice, he aims to situate hermeneutical approaches as valuable tools within the given TA process, at the same time and in order to deflect an unreflective hunger for the future, he argues more foundationally for hermeneutic TA. In the meantime – and this takes us back to the beginning of our conversation – we all seem to agree with George Khushf's critique of the form of technology assessment that takes as its starting point the separation of science, technology and the humanities, as well as the separation of epistemology and ethics. If we look closely enough, we are already there – already where Science and Technology Studies or the European Commission's "three Os" of "Open Science" seek to take us. However, though there is an awareness of this among groups like us, most of the work continues as if these separations were given. As Sascha Dickel pointed out, once we appreciate what other actors are doing and how they are thinking, hermeneutic TA cannot be anything very special but a sort of reflection or reflexivity that is embedded in ongoing practice.

Notes

1 The following provides a reconstruction of central topics at a workshop in Karlsruhe (April 26 and 27, 2018). It was assembled and edited by Vera Borrmann, Christopher Coenen, Alfred Nordmann. There were some written inputs to the workshop. Extended extracts from these appear as indented quotations, often prompting the subsequent discussion which started with a consideration of Nordmann and Grunwald, this volume chapter 2. - Not all workshop participants atttended it for its entire duration.

2 In short, he analyzed the electrical sparks of hundreds and later thousands of neurons that carried motor messages, converted them into digital signals and sent them to an artificial devices (robotic, mechanical, computational) that reproduced the voluntary motor wheel of that brain, and at the same time the artificial device sent back sensory signals to that brain. In his experiments with monkeys, he could show that the brain of the monkey in experiment adopted the artificial device and could move it with that feedback loop according to its own will (compare e.g. his TED Talk 2012: *A monkey that controls a robot with its thoughts. No, really.* https://www.ted.com/talks/miguel_nicolelis_a_monkey_that_controls_a_robot_with_its_thoughts_no_really#t-218368, checked September 29, 2021).

3 Jameson (2007) goes on to say: "It is this present moment – unavailable to us for contemplation in its own right because the sheer quantitative immensity of objects and individual lives it comprises is untotalizable and hence unimaginable, and also because it is occluded by the density of our private fantasies [...] that [...] is offered to us in the form of some future world's remote past. SF thus enacts and enables a structurally unique 'method' for apprehending the present as history, and this is so irrespective of the 'pessimism' or 'optimism' of the imaginary future world which is the pretext for that defamiliarization" (288, compare also 232).

4 The film is based on the novel of Kazuo Ishiguro *Never Let Me Go* (2005).

References

Anders, G. 1980. *Die Antiquiertheit des Menschen Bd. II: Über die Zerstörung des Lebens im Zeitalter der dritten industriellen Revolution.* München: C.H. Beck.

Arendt, H. 1989. *The Human Condition.* Chicago; London: The University of Chicago Press.

Callon, M., J. Law, and A. Rip, eds. 1986. *Mapping the Dynamics of Science and Technology: Sociology of Science in the Real World.* London: Macmillan.

Carlson, R. H. 2010. *Biology Is Technology: The Promise, Peril, and New Business of Engineering Life.* Cambridge, MA: Harvard University Press.

Eidinow, E., and R. Ramirez. 2016. "The Aesthetics of Story-Telling as a Technology of the Plausible." *Futures* 84, 43–49.

Habermas, J. 1971. *Theorie und Praxis: Sozialphilosophische Studien.* Frankfurt am Main: Suhrkamp.

Jameson, F. 2007. *Archaeologies of the Future: The Desire Called Utopia and Other Science Fictions.* London; New York: Verso.

Jasanoff, S. 2016. *The Ethics of Invention: Technology and the Human Future.* New York: W.W.Norton.

Jasanoff, S., & Kim, S.-H. (2015). Dreamscapes of Modernity: Sociotechnical Imaginaries and the Fabrication of Power. 2013.

Jonas, H. 1979. *Das Prinzip Verantwortung: Versuch einer Ethik für die technologische Zivilisation.* Frankfurt am Main: Suhrkamp.

Levinas, E. 1961. *Totalité et infini. Essai sur l'exteriorité.* La Haye: Nijhoff.

Nordmann, A. 2014. "Responsible Innovation, the Art and Craft of Anticipation." *Journal of Responsible Innovation* 1 (1): 87–98. 10.1080/23299460.2014.882064.

Nordmann, A., Grunwald, A. 2017. "Hermeneutic Technology Assessment: Why It Is Needed and What It Might Be." In *Contours of the Future: Technology and Innovation in Cultural Context*, edited by Kuznetsova, D.I., Sergeeva, V.V., Almazova, N.I., Nikiforova, N.V., 21–25. Saint-Petersburg: Asterion.

Rip, A. 2009. "Technology as Prospective Ontology." *Synthese* 168: 405–422. 10.1007/s11229-008-9449-9.

Rip, A. 2012. "Futures of Technology Assessment." In *Der Systemblick auf Innovation – Technikfolgenabschätzung in der Technikgestaltung*, edited by M. Decker, A. Grunwald, and M. Knapp, 29–39. Berlin: Edition Sigma.

Robinson, D. K. R. 2008. "Complexity Scenarios for Emerging Techno-Science: Addressing Strategy-Context Fit by Prospecting Level Dynamics of Governance." *Third International Seville Seminar on Future-Oriented Technology Analysis: Impacts and implications for policy and decision-making – SEVILLE.*

4 The Questions of Hermeneutic TA – Towards a Toolbox[1]

Martin Sand

Hermeneutic TA and Its Objects

With the concept of hermeneutic technology assessment (TA), Armin Grunwald proposes a third mode of assessing emerging technologies (Grunwald 2014b), aside from a first mode that is predictive and quantitative and a second mode that can be participatory, qualitative and which considers a "corridor of sensible futures." Before I will suggest some programmatic additions to this third mode of orientation in the form of a list of research questions, I will start by briefly examining hermeneutic TA's methodological foundations.

As the term "hermeneutics" in this composition indicates, hermeneutic TA is concerned with understanding. This immediately raises the question, what is the object of understanding in such approach? What shall we understand as technology assessors? Emerging technologies? Strictly speaking, there is no such object as "the emerging technology" that could be assessed. The predicate "emerging" signifies that the technology is incomplete, in flux, in the process of being development, and yet unfinished.[2] Depending on the stage of emergence, these technologies might exist as mere technical drawings, as circumscriptions in funding proposals, as buzzwords in advertisement headlines, or as prototypes, in which case they would have found their clearest material manifestation (see also Dickel in the present volume). Clearly, however, even in such a stage as prototypes, assessments of the emerging technologies are highly fragile and uncertain: Relevant social and ethical consequences that arise only through societal use (and potential abuse) of those technologies, or their widespread diffusion (individually induced minor harms that collectively cause collateral damage), remain hard to predict (Collingridge 1980). Not only for an institution such as TA, which is committed to the task of providing information and evaluative orientation for policy-making (Grunwald 2019), this is a shattering insight. Also, other societal institutions and stakeholders are challenged by this: Surely, few of us want to become part of a social experiment with unpredictable and

DOI: 10.4324/9781003322290-4

possibly devastating outcomes from which one cannot withdraw and one has not consented to participate in (van de Poel 2011).

The release and distribution of emerging technologies without previous assessment, sufficient supervision, regulation, or, more generally, suitable policies would seem to constitute such irresponsible social experiment. Thus, what else – if not the emerging technology itself – is there to understand and assess to provide guidance for this task? It is the pre-diffused, possibly non-material manifestations, visualizations, articulations, and verbalizations of the emerging technology that hermeneutic TA is homing in on that provide the research material that hermeneutic TA aims to study (Grunwald 2020).

The first reason we have encountered for doing this is the inaccessibility and inexistence (or not-yet-existence) of the emerging technology itself and the fact that such manifestations serve as suitable proxies for an early assessment. The second reason for doing so, which is related to the first, is that those early visualizations and verbalization of a technology can be seen – if the semantic stretch is allowed – as a trace, if only a blurry one, to their future (Sand 2018b).

Usually at the point, when TA recognizes the public impact of a technological future, the future has already gained significant traction in society, attracted public attention, and its momentum is further increasing (Nye 2006). While there is a clear significance in this moment of the societal attribution of meaning to a technological future (Grunwald 2017; van der Burg 2014), the pathway to its realization even at this point is not certain or determined.

This observation needs qualification and shall be examined briefly because it also entails a threat for the hermeneutic approach: Is there not a danger, one might ask, that hermeneutic TA is directing its attention to a technological future in the present that will never materialize, providing futile orientation while missing out on developments that impact technological change much more profoundly? Before even getting started, must hermeneutic TA not implicitly endorse a mode 1 assessment of a certain technological vision, predicting the likelihood of the vision to materialize, to determine whether a mode 3 assessment is actually required and justifiable in the respective case? Without such assessment, one might vainly provide orientation for technologies that are doomed to fail. One can delineate at least three ways in which the connection between a technological future and its realization can be severed or present us with a misleading starting point for an assessment (keeping in mind that there are numerous others, which could be mentioned here):

a) Obviously, some technological futures simply fail to materialize for various social or technical reasons. Thus, no matter how hyped and how much significance is ascribed and public attention granted; not every technological future turns into reality – some don't even get close to that.

b) Many social developments might actually have a bigger impact on technological change than the debated and hyped technological futures on which hermeneutic TA sets its eyes, as I have outlined elsewher (see Sand 2019).[3] Consequently, shouldn't those developments receive more attention in a hermeneutic TA approach?

c) Lastly, some technologies materialize without having ever been envisioned or planned in a forward-looking manner. They come out of the blue and appear suddenly and quasi-accidental. Hermeneutic TA will be caught as much off guard as policymakers and other stakeholders. Taken together, do these possibilities not seriously threaten the meaningfulness of focusing on technological futures to provide orientation?

Having introduced these distinctions, we are now equipped to respond to each of these possibilities and defend hermeneutic TA against the charge of futility and stabbing in the dark. A focus on technological futures to provide normative orientation can be justified despite those possibilities. Focusing on technological futures that are publicly ascribed with meaning and that receive attention presents us with the biggest chance to yield relevant knowledge to orient policy-making.

Option c), while being clearly possible, is a rare occasion. In every stage of technological change, human decisions are being made (Johnson 2015). Technologies never assail societies without being frequently pushed and steered by various human decisions. In those rare occasions, where such process is superfast and almost sudden, no evaluative discipline or method can cope. A technology that appears suddenly, or by chance, escapes any sort of preemptive assessment. Therefore, this possibility does not undermine hermeneutic TA, it presents, in fact, a challenge for all sorts of preemptive assessments. A general instantiation of the precautionary principle as a meta-principle to regulate socio-technical change and slow such processes down is the only and most controversial response to this, which exceeds, however, any particular process of concomitant TA and must be based on a more general view regarding the acceptable risks that can be imposed on society without knowing more about the risks of a particular technology. Regarding option b), hermeneutic TA, I believe, has to acknowledge on the one hand to a certain degree that it has, despite its transdisciplinary aspirations, certain boundaries: Its research scope cannot cover all possible sorts of (global) social dynamics that might have an impact on technological change. However, there is no reason that hermeneutic TA could not also try to open itself up to accommodate and look more broadly into those futures that initially appear to lack technological content and that seem at first sight mereley "mundane" given their potential to nevertheless redirect socio-technical change (Sand 2019). Such opening up stands in no contradiction with any of its other methodological convictions. Lastly, there is the possibility a) that technological futures are being assessed without materializing. The orientation and normative guidance that have been provided appear to be futile and resources have been wasted. That view, however, rests on a misunderstanding: First, hermeneutic TA's intent of gaining knowledge clearly aims at – but also goes beyond providing orientation for policy-making. For instance, if we better understand the human condition based on an assessment of a technological future even if that future never materializes, we have already gained significantly. Second, in some instances, putting a halt to the realization of a technological development is not a lamentable development that indicates a waste of TA resources, but is in fact exactly what hermeneutic TA after assessing the technology has recommended. Lastly, the concern presented here has its analog in the field of statistics as the problem of sensitivity rates: Without being able to expand on this in more detail, it can be said that it is *prima facie* better to assess more technological futures that never materialize than to not assess technological futures that do materialize. Hence, oversensitivity regarding technological futures seems a justifiable stance.

So far, we have established that hermeneutic TA is concerned with understanding the emerging technologies in their present guise as futuristic narratives given the absence of material objects (Grunwald 2014a). But what does such understanding entail? Which question do we have to ask, to gain a better understanding of future technologies in their manifold present representations? In the following, I will respond to this concern by discussing two examples and propose a number of research questions that might help to better understand the gravity, value, and impact of technological futures.

Autonomous Weapon Systems, Nanotechnology, and Responsibility

How can an understanding of technological futures be achieved? How does hermeneutic TA approach technological futures in the present? In order to understand how technological futures manifest themselves in the present, we should have a look at two examples; a classic and a more recent one.

Clearly, Eric Drexler's "Engines of Creation" is one of the most famous examples of a comprehensive technological vision that initiated a large public debate before there was a tangible, material technology on the nanoscale. The book written in 1986 proclaims that nanotechnology "will bring changes as profound as the industrial revolution, antibiotics, and nuclear weapons all in one" (Drexler 1986). In the book, Drexler envisions life-prolonging molecular machines that would journey through the human body and repair broken, agein cells infinitely, and he imagines nanofactories that absorb carbon dioxide out of the air and transform it into oxygen, thus creating an artificial photosynthesis cycle. Nanoassemblers or nanobots could be the smallest and most versatile building blocks in our economy, creating all other machines that are used by humans. It would be an exaggeration to say that the "Engines of Creation" and Eric Drexler alone initiated the emergence of nanotechnology. There were other people, too (Smalley, Roco, and Bainbridge, etc.) (Rip and Voß 2013). But the vision provided a reference point for the community and a medium to reflect the potential of the technology and how to pursue it (Amato 1991; McCray 2013). It is fair to say, that without Drexler's vision and effort as an enactor, nano might not have gained the momentum that it got or at least only much later.

In the following second example, too, a scientist has delivered a narrative of a possible future; not to initiate but to halt or at least severely constrain a technological development through regulation. Stuart Russell writes in his book "Human Compatible" that "Autonomous Weapon Systems have been described with good reasons as the third revolution in warfare, after gunpowder and nuclear weapons" (Russell 2019, 110). A page later, while arguing that so-called killer robots are often misrepresented in science-fiction films (e.g. Terminator), he says that these weapon systems are actually not science fiction: They already exist. Surely, as an ordinary reader, one might think that Russell, who is a renowned computer scientist at University of California, Berkeley, must know what he is talking about. As an example, he mentions the Harop drone from the Israelian Aerospace Industries, and the "possible design of an autonomous weapon with a small explosive projectile" that is presented in the "Slaughterbots video." On page 111, we find a screenshot from this video that shows a hand carrying a mini-drone.

These claims deserve critical scrutiny: First of all, it is important to point out that the Israelian Harop drone is in fact "a full man-in-the-loop operation" weapon and not an autonomous drone.[4] Second, what Russell does not mention here or anywhere else in the book is that the Slaughterbot video, to which he is referring which has been uploaded on Youtube in 2017[5] and to which he refers to underscore the claim that lethal autonomous weapon systems already exist, is in fact a fictional product that he himself created and supported by Elon Musks' "Future of Life Institute." The video shows miniature drones acting as swarms, targeting individuals with high precision, and killing them effectively with little projectiles (Scharre 2017). The video has been clicked 4 Mio. times since it was released and has been widely received (Crowder 2017; Pasquale 2020). CNNTech has cut the footage and uploaded a shortened version of the clip on their own Youtube channel.[6] In his book, Russell goes on to argue that the "Swiss Defense Department has already built and tested a real Slaughterbot and found that, as expected, the technology is both feasible and lethal" (111).

This is quite an exaggeration: What the Swiss Defence Department really did was – in fact, inspired by Russell's own Youtube video – reenacting a scene from the Youtube clip by placing a 3-gram mini-explosive on a simulated head, which was effectively a gelatine-filled polyurethane sphere.

They found that the mini-explosive's "detonation caused severe damage to the object."[7] The guiding question of this singular experiment was to test "the extent of the injuries that a miniature shape charge can cause." If we shall, for reasons of charitability, let this experiment indeed count as a validation of the power of mini-explosives, this still puts only a sub-claim of the video to the test. The video proposes a host of other possibilities that – only if taken together – constitute the threat of autonomous mini-drones as presented. I will return to some of these possibilities and implicit presuppositions later. Testing whether the video's larger narrative is actually feasible would require a very different approach. In the experiment of the Swiss Defence Department, there were no flying objects involved, let alone autonomous one's that act in swarms. It is, therefore, highly inaccurate to conclude from this that autonomous drones that operate in swarm-like fashion that can accurately identify individual targets and kill them with high precision is a real possibility and it is blatantly false to state that real Slaughterbots exist. Russell has created a narrative and an own hype around a future technology that does not exist yet – and he continues to perpetuate this hype with inaccuracies that he himself inserted into the debate. Both the Swiss Defense Department and Russell pursue their own strategies entering this debate. The Swiss Defense Department states in their press release (see footnote 8) that the experiment is part of a larger agenda: "Armasuisse S+T is actively engaged in anti-drone defence. A multidisciplinary team of sensor and effector specialists are researching defensive options in conjunction with national, international and industry partners." Linking their work explicitly to the public debate about Russell's video increases awareness for their own work. Russell himself is a sturdy supporter of various NGOs that aim at putting a halt to the development of lethal autonomous weapons. Unsurprisingly, the subtitle of the Slaughterbot video reads: "If this isn't what you want, please take action at http://autonomousweapons.org/."

That this sort of activity and inaccuracy, when it comes to such public relevant issues, causes various, pressing questions about scientific integrity and responsibility is obvious. Questions regarding responsibility can be raised more generally in relation to such sorts of "visioneering" activities (Simakova and Coenen 2013). An apt description of the concept "visioneering" referring to a number of other illuminating examples can be found in Patrick McCray's, book "The Visioneers. How a group of elite scientists pursued space colonies, nanotechnologies, and a limitless future." McCray writes:

> "[V]isioneering means developing a broad and comprehensive vision for how the future might be radically changed by technology, doing research and engineering to advance this vision, and promoting one's ideas to the public and policy-makers in the hopes of generating attention and perhaps even realization."
>
> (McCray 2013)

What we have encountered previously are forms of visioneering; in the case of Russell, not with the intent to realize a future but with the intent to put one to a halt. In order to fully understand the meaning of such technological futures in a hermeneutic fashion, I propose to expand hermeneutic TA to also consider visioneers, their actions, and subsequently their responsibilities (Sand 2016; Sand and Schneider 2017). This is important, first, because utilizing unfair practices (such as spreading half-truths or exaggerations) might pose a challenge for a more rational debate about the future and the desirability of the envisioned technologies. Calling out such threats to rational debate is part of the normative orientation that hermeneutic TA can provide to improve policy-making regarding emerging technologies. Second, since such behavior could also constitute a normative reason not to trust the expertise of such agents in the future, this remark also helps to orient the debate in suggesting whose views are creditable and reliable and whose aren't. I will return to visioneering assessment in the following section when we have a clearer view on the idea of hermeneutic understanding. To summarize, it is obvious that those futuristic narratives and activities provide an interesting entry point for TA at a very early stage:

> "The factual importance and power of futuristic visions in the governance of science and in public debates are a strong argument in favour of the necessity of providing early public and policy advice in the [new and emerging science and technology] fields […]. Policy makers and society should know more about these positive or negative visions and their background. They should understand what is going on scientifically and technologically, what is or might be at stake for future developments, […], and who might be affected […]. In summary, this needs uncovering which meaning, values, and interests are hidden in the techno-futures being communicated. Thus, gaining a comprehensive understanding of the meaning of [these developments]."
>
> (Grunwald 2014a, 279)

As argued at the outset, factual importance of the vision in present discourses does not ensure the future flourishing of the envisioned technology. It means, however, that hermeneutic TA has an identifiable entry point for assessment that could at least reveal knowledge about the human condition or present interests of relevant stakeholders, and providing orientation in cases where a technological development fails is still more desirable than missing out on providing orientation for those that succeed.

Questions to Be Asked – Understanding in Hermeneutic TA

Having introduced some examples of technological visions, we should now return to the question of what it means to understand technological futures in their present guise as visions and narratives. We saw that in the absence of a material object or prototype, we are in difficulties anticipating the future impact of an emerging technologies. However, that does not mean that we are incapable of extracting other normatively relevant information that might provide orientation and increase self-reflectivity: The visions that have just been presented contain valuable raw material that provides an interesting entry point for an assessment before the materialization and diffusion of a technology. In the most common accounts of hermeneutics, understanding means grasping the meaning of something in its entirety and its embeddedness in a certain context by studying its distinct parts and their relation to another. Understanding a text in this manner could comprise an interpretation of the biography of the author, her motives and intentions, an analysis of the historical context in which the text was written and of the various parts of the text, the actions and characters of its protagonists, and so on. Thereby, a whole of the meaning of the text is being created that exceeds an understanding of its various, distinct parts. According to Jens Zimmermann, hermeneutic

> "understanding is the interpretive act of integrating particular things such as words, signs, and events into a meaningful whole. We understand an object, word, or fact when it makes sense without our own life context and thus speaks to us meaningfully. […] We have not understood a poem, for example, when we can merely repeat the words by heart; rather we demonstrate understanding when we intone the words meaningfully and are able to express the poem's ideas in our own words."
>
> (Zimmermann 2015, 7)

Thus, as Armin Grunwald suggests, to understand visions and technological futures as a whole, their meaning, and place in current debates requires asking not just a single question but a range of different questions and to investigate their interrelation. The following list has been proposed in his *Technology Assessment in Practice and Theory* (Grunwald 2018, 122):

- What do current narratives of the future signify for the relation of humanity and technology, for humanity and nature? How do they alter or transform these relations and what is regarded as being at stake in ethical, cultural, and social terms?
- Which actors are engaged in the future-oriented debates on new technology in the respective field? What are their motivations, diagnoses and observations, hopes and fears, values and interests, etc.?
- How are narratives of the future communicated and discussed among experts and the public? What roles do they play in the technological

debates or neighboring debates about future issues of our time? (visioneering assessment)

- What forms of communication and linguistic means are used and for what purposes? What extra-linguistic tools (e.g. movies, works of art, pictures) play a role in this context, and what does their use reveal?
- Is there some hermeneutic significance in the mode 1 or mode 2 orientations mentioned above? Are underlying messages or *Zeitgeist* issues part of the technology trends, scenarios, timelines, and road-maps presented?

With the intent to substantiate hermeneutic TA as a toolkit that is fruitfully applicable to various technological futures, I will in the following expand on some of these questions and notions and provide some suggestions for extending this list of questions that can be asked from a hermeneutic TA perspective, while explicating their relevance with reference to the previously introduced cases. The interesting aspect about the following set of questions is that they do not aim to predict the future of a certain technology or assess their feasibility. They side-step the epistemic problem that forms one horn of Collingridge's dilemma as introduced at the outset. However, they can nevertheless be extremely helpful in providing normative orientation. We can cluster those questions roughly in broader/anthropological questions and narrower/more specific questions that are not equally applicable to all technological futures. Broader questions intend to understand not necessarily a particular technological future, but the very fact that technological futures are indeed at all promoted and negotiated in the public realm and that meaning is attributed to them. All technological futures entail certain ideas about the relationship between technology and humans and about human nature and its possible technological transformation. Thus, looking more broadly at the societal constellation in which agents narrate technological futures could help us understand if not the human condition as suchthen at least modern men.

Broader Concerns

- An interesting question one of the broader, anthropological type that isn't trivial at all is: Why are those visions at all projecting and narrating about a **future** technology? Why do they look into the future? One might wonder: Well, why not? When looking further back into human history, we realize that this future orientation is a rather modern phenomenon. It has not always been the case that imaginations and justifications of better societies were located in the future: The early modern utopias of Thomas More and Francis Bacons "New Atlantis" presented visions of different societies that existed on earth at the same time but at another place (Bruce 2008; Kumar 1991, 57). As Krishan Kumar writes: "The early modern utopia, confronted with a still largely unknown earth, was content to find the good society on isolated islands or strange continents" (p. 57). Even earlier, some cultures looked backwards in time to seek orientation – think about Plato's references to the myth of Atlantis and the vision of Paradise in the old testament (Kumar 1991). It is only through Millenarist movements and technological change beginning in the 18th century that people started looking towards the future as a reference point where the just and the good society will be realized (Gray 2008; Nipperdey 1975). Nowadays the emphasis on "the future" is ever increasing and one wonders what this tells us about modernity.

- In which way is the belief in technological salvation metaphysical or religious? David Noble has traced religious motivations behind many techno-scientific endeavors (Noble 1999). Maybe there is a more straightforward way to understand ideas such as "mind-uploading," etc., as forms of modern spiritualism (Coenen 2006).
- Which anthropological view underlies the vision? Many transhumanists think that technology can provide a solution to a great many grievances and human miseries, including ageing and death, because they have a reductionist or naturalist idea of human nature (Sand and Jongsma 2016). Similarly, many visions of artificial intelligence (AI) imply a rather odd view about human intelligence.

Narrower Concerns

- What's in a name? Nanotechnology has always been an umbrella term for a host of different scientific methods and approaches and also a wide range of different applications (Nordmann 2006; Rip and Voß 2013). Today, nanomaterials can be found in sunscreen or paint. These different applications raise very different ethical concerns. In the visionary stage, when Drexler wrote about the "Engines of Creation" there was no tangible material object to investigate, which makes it very important to be conceptually precise and dissect what exactly is meant by a term such as nanotechnology. The discussion about AI is affected by this problem, too. AI encompasses a wide range of different technologies including symbolic AI, machine learning, and deep neural networks, each having their narrowly confined realms of application and raise their specific ethical concerns (Mitchell 2020). While machine learning deals very well with pattern recognition, symbolic AI deals much better with rule-oriented tasks. Human intelligence ingeniously combines both (and probably more that cannot be translated or pressed into any rule-oriented shape such as intuitive and emotional intelligence). Visions often narrate the future of a technological **family**, thereby glossing over the oftentimes important factual and ethical differences between their various **subspecies**.

- What is left out by the vision? This is important in two ways: First, visions always simplify the possible future realities they represent, **and** the pathways that lead to them: Technological change is complex and it is remarkable how often new technologies are promised without considering the social factors that are necessary for them to succeed. Drexler is in his book primarily concerned with technical feasibility. But, creating a working prototype of a technology is something very different, from developing a technology that is diffused and widely employed in society: If that weren't so, we would already have had videophones widely distributed in the 1980s, when they were already possible and not just now. It is the societal context and the existing socio-technical infrastructure that co-determine whether a technology can gain traction in society. We must ask from a hermeneutic TA point of view: How much have visioneers thought about the "social feasibility" in their narrative?
- Secondly, when we ask, what is being left out of the vision, we also have to recognize in which way many visions forestall answers to the difficult questions about priority setting. By focussing on a very particular concern, visions often gloss over competing concerns, which might even more reasonably quest for public attention. Patrick McCray's visioneering book contains the following striking anecdote: In August 1977, space colonization advocate Gerard O'Neill was invited to present his ideas to a thousand or so enthusiasts at the first so-called Space Day at the Museum of Science and Industry in downtown Los Angeles. While Trekkies, NASA representatives, and other stakeholders and space enthusiasts gathered inside the museum, outside the museum, as McCray notes frivolously, "they could have encountered laid-off workers waving signs proclaiming 'Jobs on Earth, Not in Space' and 'Brown, Hire an Earthling' that expressed more terrestrial concerns" (McCray 2013, 3–4). So what many visions leave out are all the competing concerns that might be even more weighty than the one presented.
- What are the presuppositions of the vision – the explicit or implicit values? This third aspect ties in with the second: Visions make (often inexplicitly) presuppositions about how society responds to a certain technology: A commentator in IEEE Spectrum does not deny the technical feasibility of Slaughterbots but points out that the narrators of this dystopian vision presumed aside from technological feasibility that: a) Governments will mass-produce lethal microdrones to use them as weapons. b) There are no effective defenses against lethal microdrones. c) Governments are incapable of keeping such weapons out of the hands of malignant agents. d) Those agents are actually capable of launching large-scale coordinated attacks with them (Scharre 2017).
- Visions usually present themselves as a solution to a problem. We must ask: Is the problem that should be solved technologically a

real problem and, if so, who has determined that it is and how has it been determined? If it is a problem, does it require technological fixing? In the UK, one can buy for a mere 949 Pounds a motorized ice cream cone, which electrically turns the ice cream on top: This fixes the "problem" that you have to turn your hand while eating ice cream. Is that a real problem that requires technological fixing? Another interesting example might be space tourism, which gained recent attention as Sir Richard Branson traveled as a first tourist with a Virgin Galactic rocket into space, thereby celebrating a stage win in the race for space tourism against Elon Musk and Jeff Bezos (Amos 2021). Space tourism advocates might claim that there is a desire for space tourism, given that there are already sales for space journeys tickets, which shows that there is a **need** for space travels. As said before, we must critically consider what the value of those desires are, if they indeed exist, compared to other possibly more urgent concerns (such as combatting climate change (Mann 2010)). But even then, isn't the desire a result of the very promise of space tourism as a nearby future technology (van Lente 2017)? Doesn't the vision create the need and, therefore, the problem to which it supposedly responds? Is there, in other words, a real problem?[8]

- We should ask: Why now? Many visions center around technologies that have prototypical precursors or resembling artefacts that precede them. The debate about cybernetics preceded the one about AI and many of the questions that are currently discussed have been discussed in similar shape or form some decades ago. If we understand these precursors, what made them fail or stagnate at a certain point and what made their promises return at another, in short, if we understand the hype cycles around these technologies, we provide context to better assess the significance of those technologies (van Lente et al. 2013). We can place the technologies into a larger historical development and see, for instance, whether the visioneer fell prey to an ongoing fashion, overestimating the potential of a certain development. As researchers and technology assessors we are not free of such influences. Stepping outside and getting a broader historical perspective, might help us to remain self-reflective and critical. If we can find out, why previous instantiations of a technology failed or challenges remained unanswered, we might be able to conclude why a currently envisioned technology will suffer the same fate.
- Lastly, visions raise pressing questions regarding the responsibilities of visioneers: Who and in which role do agents promote these narratives (Sand 2018a)? Do visioneers overstep their professional roles and responsibilities? Stuart Russell, Eric Drexler, Ray Kurzweil, and others have become major protagonists to drive socio-technical change: Their

credibility in ongoing debates and how they conduct themselves might serve as a soft indicator when it comes to assessing the objectivity and credibility of the technological futures and their risks.

Conclusions and Outlook

In the present paper, I have suggested that hermeneutic TA can serve as a proper method to assess a variety of technological futures and thereby reveal insights relevant for policy-making and tech governance. In a first step, I outlined my understanding of how hermeneutic TA responds to the epistemic horn in Collingridge's dilemma by focussing on an assessment of technological futures in the present. In a second step, I suggested that a hermeneutic understanding of such futures must interpret the various elements of those futures and their interrelation to grasp their meaning in a comprehensive manner. This requires locating them in historical context, analyzing the motifs and responsibilities of their promotors and visioneers, and assessing their claims critically.

A final part of such comprehensive hermeneutic understanding, I believe, requires that hermeneutic TA becomes self-aware and critical of its own social and historical embeddedness as an institution that aspires towards democratic legitimacy and value neutrality. This constitutes probably the most pressing concern for future research on hermeneutic TA. Gadamer writes in "On the problem of self-understanding:"

> "[…] as a hermeneutical task, understanding includes a reflective dimension from the very beginning. Understanding is not a mere reproduction of knowledge, that is, it is not a mere act of repeating the same thing. Rather, understanding is aware of the fact that it is indeed an act of repeating. […] The operation of the understanding requires that the unconscious elements involved in the original act of knowledge be brought to consciousness."
>
> (Gadamer 1977, 45)

Thus, properly understood hermeneutic TA must become conscious in which way its own unique societal position and the premises on which its agenda is built might impact how we assess the desirability and feasibility of certain technological futures and consider other non-technological alternatives viable. Can hermeneutic TA, for example, accept a future in which a certain problem that has been widely acknowledged as one is not resolved as a viable alternative to one in which it has been solved with reasonable success and without imposing risk of equal magnitude (Sand 2021)? The demand for self-reflexivity of hermeneutic TA must in future research lead to the further expansion and examination of the list of broader questions that have been introduced above.

Notes

1 The following ideas have been presented during a virtual workshop of the HYBRIDA project to which I was kindly invited by Henrik Vogt and Jan Helge Solbakk from the Centre for Medical Ethics at the University of Oslo. I am grateful for the valuable feedback from the project members and participants of this workshop. Here and there, I use "vision assessment" and "hermeneutic TA" interchangeably, since I believe that vision assessment too is primarily interested in understanding and that its objects are more than just technological visions *sensu stricto*.

2 It is doubtful, whether a technology is ever finished. Individual artefacts (cars, mobile phones, shoes) become obsolete after some time, and, thus, far no technology has ever been considered perfect. Even if that were so, future technologies create innovative pressures, change the socio-technological systems, in which the "perfect technology" is embedded, possibly undermining workability, and may more generally make us see the technology in a less favorable light. Thus, in some sense, all technology is and remains emerging.

3 In the problem of choosing, on which of the many present technological futures one ought to focus, the epistemic concern that forms one horn of Collingridge's dilemma partially returns and cannot fully be dissolved by the hermeneutic shift in perspective. If we assess technological future A for its present social significance instead of the less prominent present technological future B, we implicitly presume that social significance of a technological future is relevant for the development of a future technology. This is often, but certainly not universally true. This remains an inevitable epistemic caveat.

4 https://www.iai.co.il/p/harop (last accessed: 15.07.2021)

5 https://www.youtube.com/watch?v=9CO6M2HsoIA (accessed, 16.07.2021)

6 https://www.youtube.com/watch?v=fWzXjn5hHjs (accessed, 15.07.2021)

7 https://www.ar.admin.ch/en/armasuisse-wissenschaft-und-technologie-w-t/home.detail.news.html/ar-internet/news-2018/news-w-t/lethalmicrodrones.html (accessed, 16.07.2021)

8 The reigning paradigm in technophile circles regarding combatting health issues is to produce more "silver bullet" technologies such as antibiotics and vaccines. Given the lack of success in developing more of those silver bullets, more authors propose a shift towards "gentle medicine" (Sand and Jongsma 2017; Stegenga 2018).

References

Amato, I. 1991. "The Apostle of Nanotechnology." *Science* 254 (5036): 1310–1311. 10.1126/science.254.5036.1310

Amos, J. 2021. "Sir Richard Branson Takes Off on 'Extraordinary' Space Flight." *BBC News*. Accessed January 25, 2022. https://www.bbc.com/news/science-environment-57790040

Bruce, S., ed. 2008. *Three Early Modern Utopias: Utopia, New Atlantis and The Isle of Pine*. Oxford: Oxford University Press.

Coenen, C. 2006. "Der posthumanistische Technofuturismus in den Debatten über Nanotechnologie und Converging Technologies." In *Nanotechnologien im Kontext*, edited by A. Nordmann, 195–222. Berlin: Akademische Verlagsgesellschaft.

Collingridge, D. 1980. *The Social Control of Technology*. New York: St. Martin's Press.

Crowder, L. 2017. "As Much Death as You Want": UC Berkeley's Stuart Russell on "Slaughterbots." *Bulletin of the Atomic Scientists*, December 12, 2022. https://thebulletin.org/2017/12/as-much-death-as-you-want-uc-berkeleys-stuart-russell-on-slaughterbots/.

Drexler, E. K. 1986. *Engines of Creation: The Coming Era of Nanotechnology*. New York: Anchor Books.

Gadamer, H.-G. 1977. "On the Problem of Self-Understanding." In *Philosophical Hermeneutics*, edited by D. E. Linge, 44–58. Berkely: University of California Press.

Gray, J. 2008. *Black Mass - Apocalyptic Religion and the Death of Utopia*. London: Penguin Books.

Grunwald, A. 2014a. "The Hermeneutic Side of Responsible Research and Innovation." *Journal of Responsible Innovation* 1 (3): 274–291. 10.1080/23299460.2014.968437

Grunwald, A. 2014b. "Modes of Orientation Provided by Futures Studies: Making Sense of Diversity and Divergence." *European Journal of Futures Research* 2 (1). 10.1007/s40309-013-0030-5.

Grunwald, A. 2017. "Assigning Meaning to NEST by Technology Futures: Extended Responsibility of Technology Assessment in RRI." *Journal of Responsible Innovation* 4 (2): 100–117. 10.1080/23299460.2017.1360719.

Grunwald, A. 2018. *Technology Assessment in Practice and Theory*. London: Routledge.

Grunwald, A. 2019. "The Inherently Democratic Nature of Technology Assessment." *Science and Public Policy* 46 (5): 702–709. 10.1093/scipol/scz023.

Grunwald, A. 2020. "The Objects of Technology Assessment: Hermeneutic Extension of Consequentialist Reasoning." *Journal of Responsible Innovation* 7 (1): 96–112. 10.1080/23299460.2019.1647086.

Johnson, D. G. 2015. "Technology With No Human Responsibility?" *Journal of Business Ethics* 127 (4): 707–715. 10.1007/s10551-014-2180-1.

Kumar, K. 1991. *Utopianism*. Buckingham: Open University Press.

Mann, A. 2010. "Space Tourism to Accelerate Climate Change." *Nature*. 10.1038/news.2010.558.

McCray, P. 2013. *The Visioneers: How a Group of Elite Scientists Pursued Space Colonies, Nanotechnologies, and a Limitless Future*. Princeton: Princeton University Press.

Mitchell, M. 2020. *Artificial Intelligence - A Guide for Thinking Humans*. London: Penguin Random House UK.

Nipperdey, T. 1975. *Reformation, Revolution, Utopie: Studien zum 16. Jahrhundert*. (Kleine Vandenhoeck-Reihe). Göttingen: Vandenhoeck & Ruprecht.

Noble, D. F. 1999. *The Religion of Technology: The Divinity of Man and the Spirit of Invention*. New York: Penguin Books.

Nordmann, A., ed. 2006. *Nanotechnologien im Kontext: Philosophische, ethische und gesellschaftliche Perspektiven*. Berlin: Akademische Verlagsgesellschaft.

Nye, D. E. 2006. *Technology Matters: Questions to Live With*. Cambridge, MA: MIT Press.

Pasquale, F. 2020. "Machines Set Loose to Slaughter": The Dangerous Rise of Military AI." *The Guardian*, October 15, 2020. https://www.theguardian.com/news/2020/oct/15/dangerous-rise-of-military-ai-drone-swarm-autonomous-weapons

Rip, A., and J.-P. Voß. 2013. "Umbrella Terms as Mediators in the Governance of Emerging Science and Technology." *Science, Technology & Innovation Studies* 9 (2): 39–59.

Russell, S. 2019. *Human Compatible: Artificial Intelligence and the Problem of Control.* Viking: Penguin Random House.

Sand, M. 2016. "Responsibility and Visioneering—Opening Pandora's Box." *NanoEthics* 10 (1): 75–86. 10.1007/s11569-016-0252-7.

Sand, M. 2018a. *Futures, Visions, and Responsibility: An Ethics of Innovation.* Wiesbaden: VS Verlag für Sozialwissenschaften.

Sand, M. 2018b. "How the Future Has a Grip on Us." *Techné: Research in Philosophy and Technology* 22 (2): 262–282. 10.5840/techne201862684.

Sand, M. 2019. "On "Not Having a Future"." *Futures* 107: 98–106. 10.1016/j.futures.2019.01.002.

Sand, M. 2021. "RRI und Postwachstumsökonomie als Herausforderungen für die Technikfolgenabschätzung." In *Technikfolgenabschätzung: Handbuch für Wissenschaft und Praxis*, edited by S. Böschen, A. Grunwald, B.-J. M. A. Krings, and C. Rösch, 403–414. Baden-Baden: Nomos.

Sand, M., and K. R Jongsma. 2016. "Towards an Ageless Society: Assessing a Transhumanist Programme." In *Ageing and Technology*, edited by E. Domínguez-Rué and L. Nierling, 275–294. Bielefeld: Transcript.

Sand, M., and K. R. Jongsma. 2017. "The Usual Suspects: Why Techno-Fixing Dementia Is Flawed." *Medicine, Health Care and Philosophy* 20 (1): 119–130. 10.1007/s11019-016-9747-9.

Sand, M., and C. Schneider. 2017. "Visioneering Socio-Technical Innovations: A Missing Piece of the Puzzle." *NanoEthics* 11 (1): 19–29. 10.1007/s11569-017-0293-6.

Scharre, P. (Producer). 2017. Why You Shouldn't Fear "Slaughterbots". *IEEE Spectrum.* Filmed July 15, 2021.

Simakova, E., and C. Coenen. 2013. "Visions, Hype, and Expectations: A Place for Responsibility." In *Responsible Innovation*, edited by R. Owen, J. R. Bessant, and M. Heintz, 241–266. Chichester, UK: John Wiley & Sons.

Stegenga, J. 2018. *Medical Nihilism.* Oxford University Press.

van de Poel, I. 2011. "Nuclear Energy as a Social Experiment." *Ethics, Policy & Environment* 14 (3): 285–290. 10.1080/21550085.2011.605855.

van der Burg, S. 2014. "On the Hermeneutic Need for Future Anticipation." *Journal of Responsible Innovation* 1 (1): 99–102. 10.1080/23299460.2014.882556.

van Lente, H. 2017. *De ruimtetoerist: vragen over technologie en toekomst.* Maastricht: Inaugural Lecture.

van Lente, H., C. Spitters, and A. Peine. 2013. "Comparing Technological Hype Cycles: Towards a Theory." *Technological Forecasting and Social Change* 80 (8): 1615–1628. 10.1016/j.techfore.2012.12.004.

Zimmermann, J. 2015. *Hermeneutics: A Very Short Introduction.* Oxford: Oxford University Press.

Part II

Theory and Context

5 Technology in the Imagination of Society – A Conversation

Sheila Jasanoff and Alfred Nordmann

NORDMANN: Let's begin right at the heart of things with the term "sociotechnical imaginaries" – in contrast to that other well-established term "sociotechnical system." The latter always struck me as being somewhat lazy: "Okay, we realize that technology doesn't exist on its own. There is always some kind of social process, an integration of human beings, so let's acknowledge this by talking about sociotechnical systems." But what it refers to is simply a technical innovation, with the recognition added in that it has social implications. Accordingly, "sociotechnical imaginaries" often refer to fictional visions of so-called emerging technologies. For you, however, "sociotechnical" has a much wider resonance, and "imaginaries" go far beyond "visions." Can you say a little more about what kind of historical category you have in mind?

JASANOFF: Probably the best place to begin is by saying that for me "sociotechnical imaginaries" is one stopping point on a trajectory of theorizing, and the centerpiece of this trajectory is co-production. First of all, I've been trying to break down the notion that you first make the facts of the world, that these correspond to some external set of realities, and only then the normative kicks in and it is to some extent conditioned by the factual. On the contrary, the whole idea of co-production says that to a huge extent we're making our realities in accordance with normative commitments that order our world. I don't even want to say "prior commitments," since I think that the commitments to understanding and the commitments to living evolve hand-in-hand. In my own work – this is a sort of personal research trajectory – I have spent a lot of time on what people like to call institutions, places where at a particular point the epistemic and the normative, the value-laden and the factual become entangled and congealed. My work on political culture is about that dimension of things.

DOI: 10.4324/9781003322290-7

One of my questions is, how do societies change? We know that new things happen. We also know that hidden fissures exist that can be re-activated and made visible. If you look at the last set of German national election results (2017), it's not suddenly out of the blue that the AfD (*Alternative für Deutschland*, a right-wing party) ends up doing so well. A new mapping of the world, of the German political world, was created in 1949, and two forms of life – East and West – emerged that went their separate ways. The fact that 1989 and 1990 happened and that these two forms of life were reintegrated does not mean that all of those generation-long commitments that were built into practices simply disappeared. They were submerged. If you look at the election results, you see a political map reemerging that had been rendered invisible. Here, imaginaries are not necessarily utopic visions in the sense that people didn't have them before and I (Alfred Nordmann) read the correction by Jasanoff as follows ("them" seems necessary) and now have concocted them as something: The building blocks may have been lying around disused and are getting put together again in certain familiar ways.

Sociotechnical imaginaries to me are an extension of co-production, making the basic point that nothing we do in our contemporary lives fails to walk through the regions which we define as science and technology. It is not the case that there is a "Habermasian" communicative sphere that is sitting there, or a "Rawlsian" public reason, and then there is technology. I would rather say that however we imagine ourselves, whatever we like to do with ourselves, it is always more than just the ideas that are enabled or powered by technological possibility. Technology is also, at the same time, modulated by its public setting and answerable to ways that people wish to live. Then it becomes a research topic in and of itself that you can always query new technological ideas as to how they are associated with seeming changes in the directionality of society. Charles Taylor is in a sense the social equivalent of Thomas Kuhn. I mean, he's describing social paradigm shifts. I don't entirely agree with either of them, since it is not a shift so much as a crystallizing of another possibility, raising the question: What provoked or made it possible that these things came into view and crystallized in this way at this time?

NORDMANN: The crystallizing of possibility would be the moment when something appears. There is this concept of real possibility according to which something becomes salient in a historical situation, from a historical point-of-view, from a condition in which society finds itself.

JASANOFF: Salient and sufficiently achievable such that people start working towards it, and then it stops being science fiction or one person's idea in the back of their heads. There are all kinds of scattered ideas floating around, and then somebody can realize there are ways of materializing one. There's nothing predictive about it, but you can make sense of it – at least historically.

NORDMANN: One example of this might be the city of Lübeck around the year 1900. It was a town of merchants and traders that began envisioning itself – late in the game – as moving into the industrial world, refashioning itself by building a large foundry and creating a new working class of laborers housed in a company town. Here we have citizens re-visioning social relations, their notions of modern subjects and only incidentally, their future (see Behneke).[1]

JASANOFF: That's really closely related to one of the bigger projects I'm wrapping up at the moment, which is on the reception of the *Silicon Valley* imaginary in a number of different cities. This has, of course, assumed a kind of inevitable quality: Everybody wants to be *Silicon Valley* or the next cradle of a *Silicon Valley*. It is a kind of industrial revolution of the moment, and then all of these cities that are being like Lübeck decide that this is the self-image they want to latch on to at this time. With Sebastian Pfotenhauer I have worked on another story like this, that of having your own MIT in different places all over the world (see Pfotenhauer and Jasanoff 2017).

One of the things that people miss when they work with this idea of imaginaries is that they really need to pay attention to the aspirational quality of those visions. And that means, also taking into account the flipside – the diagnosis of a deficit that needs to be remedied. What all those cities have in common that suddenly wanted to have their own MIT was a self-diagnosis of almost a kind of illness: What is it that ails us? And of course, it was perceived as an economic ailment, a foundational ailment: What do we need to do to repair our productive capabilities? If I was developing that set of ideas further, I would bring it into line with thinking about the body as an economy, which of course has a long history as a productive system, and of reflections on its limits.

We also talk about the body politic and so we say that those cities are great incubators for new forms of life. With utopic and dystopic visions the diagnostic comes into play, suggesting polar ends of a spectrum of possibility (see Pfotenhauer and Jasanoff 2015). But there is a middle ground on which many more people are dealing with certain satisfactions and dissatisfactions. In a sense, trivially, everybody has always got some ailment. You get up in the morning and say, "Today I have a headache, or why do I really feel so arthritic, or I ran too much yesterday." Interestingly, this doesn't usually build into a call: "Therefore, I must regenerate." Imaginaries, however, are not just localized sensory perceptions that something is the matter. It's not like you've been sitting too long in a chair, and you need to move the body. Imaginaries come into play when suddenly you realize, "I have spent my entire life sitting in a chair, and now I need to do something completely different."

NORDMANN: What you say about the body as an economy with limited productive capabilities, brings to mind the discussions regarding transhumanism. In this discourse, too, one often comes upon the self-diagnosis of foundational ailments. There appears to be a felt predicament primarily of men who basically hate their bodies and thus imagine forms of reproduction that wouldn't involve women. Someone like Nick Bostrom is actually very explicit, in a deeply troubling way, about the many respects in which he finds himself deficient and lacking (Bostrom 2008).

JASANOFF: I haven't seen a serious gender analysis regarding this issue. I don't know, for instance, whether transhumanism for men tends to be more of the kind of singularity idea, of brains uniting with other brains, whereas for women it's changes of the imperfect body or something like that.

NORDMANN: I wouldn't claim that I made an analysis of it, but what I often read is a description of how miserable or disheartening it is to imagine that some wonderful brain has to sit in this very deficient kind of body.

JASANOFF: And die with it.

NORDMANN: And die with it, indeed. So let's take this a step further to the question of whether or to what extent these imaginaries are anticipations or projections of the future. You already said that they're not predictions, they're more like aspirations. That would make them more like a resolve or an acknowledgement that something has to change or that a situation has to be reimagined. There are even cases about which you might say – and transhumanism might be an example of that as well – that the imagined new way of being is actually known to be unobtainable, that it cannot be achieved. Then the imaginary really has the function primarily to articulate this kind of crisis or urgency of change.

Drawing on the notion of sociotechnical imaginaries I tried to develop this point for different conceptions of machines, seeking to show that the idea of a machine – especially astounding machines, machines that are somehow non-mechanical machines – can manifest itself in the idea of "soft machines" which Richard Jones imagined for nanotechnology (see Nordmann, this volume chapter 11). It was an exercise of imagining an unheard-of, happy, and harmonious integration of material things, nature, and people. Even from Jones's own perspective this was utterly fantastic, but therefore or nevertheless became a socially utopian idea: With soft machines come notions of soft governance, a non-coercive form of creating win-win situations where everything turns out to be right.

These are very rich imaginations for something of which everyone knew or could have known that it cannot be technologically realized – Richard Jones wrote as a physicist, fully aware of the limits of technology.

JASANOFF: What was a soft machine? It's not a nano implant in the body?

NORDMANN: No, that is exactly the point. Jones said the "nano-visionaries" of his day are talking about nanoscale implants in the body as if they were scaled-down machines. He began by arguing that these visionaries didn't know what they were talking about: If you were to imagine a machine at the nanoscale it would have to work on radically different principles because a little motor couldn't overcome the forces that act at the nanoscale. You have to have a machine that is somehow integrated into the system in which it operates. It would have to be more like a flagellum – which is actually not a machine at all, you might say. By using the metaphor of a machine, Jones thereby reimagines what a machine might be. And it is thus that he became the official ombudsman for nanotechnology in England, conducting various soft governance experiments with public engagement. Even though he was a physicist and came up with the imaginary of soft machines – seemingly a question of materially realized technology – these machines insinuated socially utopian ideals of the way in which humans and technology can be happily integrated.

All of this, in some sense, is based on a "what if" with a perfectly unreal premise. I don't know whether we need to go into the details of it, I proposed it only to exemplify that these imaginaries don't have to posit a future. They might arise just as we become aware of the possibility that things could be different, very different, indeed.

JASANOFF: But to me, that account is still too much focusing on the possibilities of science and technology. I'm just talking about the description you've given me so far because I'm sure that that's not where it ends. The correlative dimensions of the pieces of the normative order you try to achieve depend on what kind of human subject and what kind of social connection we are dealing with. Not just the so-called "softness" and free-standing ideas of the interplay of things.

What my collaborators and I are showing is that whatever narrative is employed to imagine new possibilities, one thing that tends to remain constant in the Silicon Valley imaginary is that the new order is still imagined as a kind of market, a market order. I'm reading a very nice chapter on San Francisco written by Brice Laurent and a colleague of his (see Laurent and Talvard 2017). They do very well at describing how post-Silicon Valley San Francisco has constituted itself as a city of permanent innovation, so everything is open for re-imagination all the time. But, I think what they missed – and they get the fact that democracy itself is a site of innovation – is the conversion of the political subject into, in effect, a consumer subject, so the public good becomes a matter of economic benefit. All of the projects are keyed to some notion of welfare improvement that already has an economist framing behind it. For example, two-sided park benches – you can say that it enhances public space, but it's a consumer product, not really dealing with the publicness of public space except in economist terms. In effect, it is taking already extant objects and turning them into yet more efficient sites of production. So, you already have the space dedicated to the park bench: why not make it two-sided so that more people can occupy the same space?

That's just one tiny example that leaped out at me from that chapter. I want to draw attention to a deeper level of the exercise of the imagination. To me, it is the constitutional level, that is, what kinds of subjects does the collective imagination take as fixed and given in the order of things and which ones are changeable or moveable? I think that one might then talk about revolutionary moments as ones in which those kinds of quite foundational assumptions about authority and subjecthood and so forth become overturned. The emergence of the *Citoyen* as a figure in the French Revolution can be seen as a revolutionary moment because feudal and post-feudal ideas of the right relationship between layers of hierarchy – top and bottom – became overturned, and people started understanding social change in a radical way. But most moments of technological innovation are not foundationally significant in that sense. I would like to stop talking about Silicon Valley and instead talk about contraceptives because nobody has actually written about the alteration of the gender imaginary in connection with the gaining of control over the reproduction cycle. I think that in historical terms this is more of a constitutional moment than e-mail or social media.

NORDMANN: What you said is interesting in two ways: First, in order not only to identify but also to analyze these imaginaries, it is not enough to see that they respond to a diagnosis or critique. We have to understand how they are constrained in very specific ways through the sort of things that are taken for granted implicitly, those kinds of things that are not really up for grabs …

JASANOFF: … that are not easily up for grabs …

NORDMANN: Exactly. And this introduces secondly a kind of normativity that does not consist in saying what is right or wrong. You point out how these imaginaries themselves and their production are conditioned by deeply rooted normative conceptions. It is very difficult to question the logic of the consumer subject as one tries to imagine a better use of a park bench, or even as one re-imagines what a public space might be. So, these notions get interestingly intertwined.

JASANOFF: A lot of the imaginaries stay in place because of the extreme power of Bourdieu-like conditioning of the self, at bottom, to accept certain things. We see that a coming to power of a rogue or rebel –

whatever you want to call it – reveals the weaknesses in the structure, and so something that seemed solid becomes atomized again. What we've been seeing in America especially during the Trump years is just how much of our political order is made up of tacit rules. Once we changed the valances of good and bad in that tacit regime, why shouldn't the leader of the free world be exchanging love letters with a murderous, brutal thug in North Korea who chooses to eliminate all his opponents? It's as if the polite social order doesn't even have words any more to speak about these unimaginable things. You have a change in the imagination and a change in the ground rules for what is considered to be feasible or not feasible, or maybe you don't have a change at all. We will see whether those tacit norms have a kind of tensile strength that survives these attempts to break things up.[2]

NORDMANN: You also said that instead of always having to talk about Silicon Valley, you would like to talk about other questions such as contraception as another arena in which technology might have, in some sense, altered our conception of the subject, of who we are, and doing so even more profoundly than Silicon Valley ever will. But, it seems like you're also constrained, or you feel constrained in terms of the difficulty of drawing attention to some domains.

JASANOFF: There is a theoretical point in what I want to say. Science and Technology Studies (STS) makes a profound analytic error in choosing to do what Latour proposed as a "Rule of Method": Yes, you can follow the scientists, but all you're doing then is you're re-inscribing the fetishistic qualities of the moment. In order to understand society you have to have science importantly in it, but not: "Give me a laboratory and I will raise the world." I will raise the world and the laboratory is one instrument to do it with. Sometimes I like to say Actor Network Theory (ANT) has proved to be methodologically very rich and so everybody does it, and now they're all talking about new materialisms. But, the age-old questions of meaning – I think both religion and political theory wanted to grapple with those – are: Why is there suffering? Why is good unrewarded, and why is bad more successful than it should be? What does ANT tell you about any of those things? If you want to answer them, you have to bring in other considerations, such as an understanding that the social order is also a political and moral order. There are things worth fighting for and things that are relevant to collective and public goods, and hence worth defending.

We have evolved certain ways of dealing with these in theoretical terms, and one of the things we have evolved is theorizing what a political subject is, but a political subject today is one that is tied in all kinds of ways to the products of our ingenuity. One of the ways in which people imagine the human subject – political, social, or whatever – is in relation to technologies, but that's not the be-all and end-all of how we imagine subjectivity. We imagine a constitutional subject, we imagine an educable subject, and of course an economic subject. The subject comes in a variety of flavors, and only when we perceive the layered quality of subjectivity should we start thinking about the role of science and technology in propelling some of those identities forward, possibly at the expense of others.

The other day Charles Taylor was delivering the Walter Benjamin lectures. He called attention to the fact that the French revolution is the grandmother of social imaginaries: *Liberté, égalité,* and *fraternité* are the three pillars of liberalism, and today we seem to have lost sight of *fraternité*. Instead, we are only thinking of the other two.

Égalité evokes a deep question of who is in the equivalence-making domain in the first place. The American constitution shows that it was white males, with accent on white. And actually, it was easier to argue for black males to be emancipated than for women of any sort to be emancipated.

And then there are the moves by which subjectivity itself became enlarged. If you're an STS scholar, you ask, "Well, what roles were played by science and technology or scientific and technological possibility in creating the 21st century subject, and who is that subject anyway?" But here STS has to reflect on the standpoint of that question - what is it that we want to elucidate? Though this becomes foregrounded very often, it is not about how a technology gains hold in public imagination. Yes, that is a worthy question and it should be investigated, but, coextensively at the same time, and more significantly: What is the imagination of society in which the wheels of technological progress, the very idea of progress itself, are being reimagined? This is that deeper level of the exercise of the public imagination.

You find that there was this period in the 1960s when control of the reproductive cycle became a thing, because of contraception, and this leads you to look at today's gene editing efforts, which also involve control of the reproductive cycle, with completely different eyes. It also makes you look with different eyes at the American assaults on abortion because you see how imperfect and incomplete that control over reproduction actually was. In effect, you see the religious Right's reinvention of the fetal person as an entitled subject to offset against women's "liberation" from pregnancy, and fetal technology and the science of fetal pain then become part of a bigger story in US politics. A story of how difficult it is to maintain a common moral world, in which people see that biology still plays a role in subject formation, that it still takes nine months to gestate a baby whose status in that time neither law nor science have fully determined, and this ambiguity allows for both political and technological interventions with untold consequences for the economy and for society.

NORDMANN: When Armin Grunwald and I started talking and the notion of hermeneutics came to the fore, it first appeared that the main focus would be on how we talk about ourselves when these imaginaries come into play. I have been arguing, that this is not enough. If you want to call it "technology assessment," there has to be some sense in which these stories about imaginaries and the constraints that shape them are also stories about the work it takes to change ways of thinking, how hard it is to integrate new ideas with tradition and an established social order. In our short text about hermeneutic TA (see Nordmann and Grunwald, this volume chapter 2) we thus talk about a sensorium of sorts: We have structures of law, of science, of thinking, of biology that resist transformation and that prevent all of reality being resolved into a new vision. Even when we imagine futures, even when we imagine ourselves as different, or even when we imagine a German Silicon Valley, we are, at the same time dealing with irreconcilable notions that indicate the precariousness, even futility of some of those aspirations. Without being normative in the extraneous sense of applying general principles – how can we get a sense of what works, how are working orders or social orders brought to their limit?

JASANOFF: One question is, to what extent is there an obligation or a methodological necessity for people to be reflective about themselves? As a prescriptive matter, I don't want to be like the technology assessor from the past whose job was to tell the rulers: This is what you should do. But, as you know, even the Office of Technology Assessment (OTA) decided that that was not its best approach, but instead it should try to present alternative scenarios and options. The OTA's institutional position was fairly well discernible from the ways it formulated its reports, but nevertheless *officially* it said, "Here are options one, two, and three, and here are the costs and benefits associated with each." How to act then was ostensibly left to the policymaker.

This posture was still an expert posture in the sense that OTA had to select the scene and objectively identify these conditions, these possibilities, and assess the costs and benefits by taking note of certain factors, thereby excluding others. For today's analyst the whole project of critique, and what she chooses to take on as objects of critique, to point our collective attention to, is a moral and political commitment. This commitment dictates the method, among other things, or at least, it is intimately tied to the method.

In my case, looking at how people are talking about their pasts, presents and futures is not quite enough without looking at who those people are. That's why I'm interested in the comparative level that opens up across relatively stable social orders such as those of modern nation-states. There may be many alternative imaginations floating around inside nations, but even then they would be constrained by certain cultural preconditions. When someone like Konrad Ott describes his citizen juries and says that in the end its members were all just very skeptical about their own expertise (see Merk et al. 2017), I have to wonder whether that's because your average German citizen off the street just isn't conditioned to be radical in their self-imagination or whether the selection process left out the real radicals.

What would happen if you were doing this in a country with a truly radical or more radical environmental movement? Brice Laurent was describing the anti-nano mobilizations in Grenoble (see Laurent 2017). The activist group there was "French" enough to recognize that just being in the same language game as the government might lead you to defeat. Therefore they talked non-sense and just uttered non-sense syllables while ostensibly participating in the process. You can argue that they still respected the process, thereby ratifying some ground rules, but nevertheless that's a very different reaction from the types of deference that Ott's five citizens juries displayed. Why does it happen that this particular imagination of "how to be disruptive" makes sense in France, but would not make sense in Germany?

Back in 2007, I remember we had an exceptionally productive meeting of the Science and Democracy Network, and one of the conclusions from that meeting was that the critical aim of public participation is already to some extent lost if it becomes a formal requirement: Once it's ruled by government agencies, they just say, "These are the ways that you shall participate if you choose to." As soon as you become "proceduralized" in that way you are losing some of your freedom of expression.

To me one of the canonical examples of creative critique is Gandhi and the Salt March. The Salt March was one of these great liberationist movements in Indian anti-colonial history. Gandhi was educated in Britain. He was a lawyer, practiced in South Africa for a while, and the early pictures of him are wearing a black suit and cravat. Then a transition happened. The loincloth-wearing Gandhi is a native Indian figure, a figure that keyed into existing imaginations of sanctity and being "for the people," not in South Africa but in India. The British were producing, manufacturing salt on industrial scales, and then charging people for it. There was a salt tax, but Gandhi led thousands of people on this march to the sea to show how you could take seawater and extract salt out of it for free. It was one of those breakdowns of order, but it was also a hermeneutic TA mode, that is, avoid this technology, and go for that other technology, in order to achieve the same practical results but with a radically different politics. This was also one of the moments I was talking about before. There was a possibility for action because the social order became porous. You could find the empty spaces to reimagine or rethink it, and these spaces involved technology.

I learned an enormous amount from a sixties radical Dutch friend of mine who said to me, "Why are you so interested in creating STS at

Harvard?" I persuaded him that Harvard itself is an imagination that has colonized the world firmly enough to have ready-made power, and if I succeed there, I would have its force behind me. He was never so directive, but he pointed out that my thinking about Harvard was flawed and that I should think about, in effect – not his word, my word – the porousness in the university system. I accepted his analysis. I ended up building the Science and Democracy Network, and also the STS Program at Harvard, using spaces inside the structures of the university that other people had left lying open and fallow. In that respect STS at Harvard was a kind of squatter culture.

The question is, how can such a squatter culture insinuate itself into the dominant narratives of technology, like the stories of garage tinkering that eventually generated the vast earnings of Silicon Valley? You wouldn't easily get to that story if you started with the software that those engineers built – I think you'd have to begin somewhere else. The methodological point is to get to the full picture of the potentialities of reform or transformation or progress in a society. Don't begin with the technology, do not follow the scientists or the engineers, rather follow the aspiration and the rebellion – or maybe do not follow the rebellion. Concerning women's rights and the question of equality, for example, maybe what you follow are the lawyers who argue for maintaining *Roe v. Wade*, because they were trying to hold on to a beachhead in an attempt to achieve a kind of equality that transcends the biological. It doesn't do away with the biological, I think this would be a false move, but it says that our job should be to use everything that makes equality itself sensible as a principle and then lets us fight for that. This is today a seemingly conservative stance, but the ongoing story of *Roe v. Wade,* which a conservative Supreme Court overruled in 2022, keeps exhibiting the porosities in the practice of reproductive rights. The struggle over Roe is a necessary instrument for maintaining an ongoing sense of what gender equality needs to look like, not as a goal that can be achieved in final form but as a constant aspiration.

Notes

1 Timm Behnecke's dissertation research at Technische Universität Darmstadt and the Zentrum für Kulturwissenschaftliche Forschung Lübeck.
2 Ironically, perhaps, a 77-year-old-guard politician was charged to provide this strength.

References

Bostrom, N. 2008. "Letter from Utopia." *Studies in Ethics, Law, and Technology* 2 (1): 1–7.

Laurent, B. 2017. *Democratic Experiments: Problematizing Nanotechnology and Democracy in Europe and the United States.* Cambridge MA; London: The MIT Press.

Laurent, B., and F. Talvard. 2017. "*Real-time Democracy: Imagining the City of Permanent Innovation,*" *i3 Working Paper Series*, 17-CSI-01.

Merk C., G. Klaus, J. Pohlers, A. Ernst, K. Ott, and K. Rehdanz. 2017. "Public Perceptions of Climate Engineering: Laypersons' Acceptance at Different Levels of Knowledge and Intensities of Deliberation." *GAIA* 28 (4): 348–355.

Nordmann, A., and A. Grunwald. 2017. "Hermeneutic Technology Assessment: Why It Is Needed and What It Might Be." In *Contours of the Future: Technology and Innovation in Cultural Context*, edited by Kuznetsova, D.I., Sergeeva, V.V., Almazova, N.I., Nikiforova, N.V., 21–25. Saint-Petersburg: Asterion.

Pfotenhauer S., and S. Jasanoff. 2015. "Imaginaries of Innovation: Political Cultures, National Visions of Development, and the 'MIT Model' in Four Global MIT Partnerships," *Workshop on the Changing Political Economy of Research and Innovation.*

Pfotenhauer, S., and S. Jasanoff. 2017. "Panacea or Diagnosis? Imaginaries of Innovation and the 'MIT Model' in Three Political Cultures." *Social Studies of Science* 47 (6): 783–810.

6 On "Not Having a Future"[1]

Martin Sand

Hermeneutic TA, Meaning and Public Awareness

Technology assessment (TA) has a social function: It is an institutionalized, problem-oriented research endeavor (Grunwald 2015, 66; 2019, 52). Its research follows the societal demand to advice, inform, and thereby enhance political decision-making regarding emerging technologies. By assessing emerging technologies' potential effects and social repercussion, TA provides orientation for stakeholders, policymakers, and the public. As Armin Grunwald rightly argues in his paper "Modes of orientation provided by futures studies: making sense of diversity and divergence," these requirements are often hard to conform to regarding the (increasing) amount of far-fetched visions of technologies and visionary discourses that surround some currently emerging technologies such as synthetic biology, in-vitro meat, and AI (Grunwald 2014a, 7). In such cases, the non-existence (not-yet-existence) of concrete artifacts and prototypes makes modeling and prospection of these technologies' futures virtually impossible. Nevertheless, to master the challenge posed by the Collingridge-Dilemma, it is advisable to direct the attention on these technological futures *in the present*. Furthermore, there can be no doubt that those visions already have an impact in the present. They affect societal discourses, they are attributed with meaning, and thus develop significance for technological development and policymaking: "[T]he assignment of meaning is the most upstream point reachable in public and ethical debates on NEST [new and emerging technologies]" (Grunwald 2014b, 279; 2017, 101). Because of this, Grunwald suggests a "third mode" of accessing the future or, in other words, orienting towards the future: A hermeneutic extension of TA (Grunwald 2014a, 7; 2017). By scrutinizing and understanding the emergence of those futures and their "immanence in the present," we can learn something about the here and now (Grunwald 2012, 270). This suggests a third, "non-consequentialist," way of studying the future, which also leads, according to Grunwald, to insights that can provide orientation for policymaking, albeit of a different nature (Grunwald 2015, 66).

DOI: 10.4324/9781003322290-8

Following up on these suggestions and in order to contribute to the bourgeoning – and in my mind reasonable and thoughtful – idea of a hermeneutic TA as presented before, I will ask in this contribution, which kind of knowledge can be acquainted from such perspective and whether this kind of knowledge can be utilized in policymaking. What can we learn from investigating the present(s) of currently emerging technological futures? In order to answer these questions, it will be helpful to first shed a retrospective and critical light on the recent past of TA's engagement with technological visions and scrutinize which kind of visions and how they were approached so far by TA. Hence, in the present chapter, I propose a truly hermeneutic stance on the visionary research previously executed in the context of TA and reflect on its implications. This will reveal that a small group of highly educated, male, well-off people from the Northern Hemisphere has promoted the visions that attracted most of the attention of TA and the public discourse on our socio-technical future. Acknowledging this should make us simultaneously aware that there are numerous people who neither entertain their own vision of a desirable socio-technical future (for a variety of reasons, which I shall explore later on) nor do they have the means to strategically position them in the discourse and, thereby, contribute to their realization. We might be inclined to speak of those people as "not having a future." I will investigate the meaning of this ambiguous notion and try to scrutinize the origins and normative implications of this misbalance. Instead of providing a concluding account of the trope of "not having a future," this chapter aims at initiating an important discussion about inequalities in future making. The following insights should remind us that meaning is often not only affected by the content or design of a future or vision, but also by the social standing of the agent who promotes it. In this manner, my contribution wholeheartedly breathes the spirit of Future Studies (FS) in presenting a "systematic mode of critical enquiry" concerning the inequalities of future making (Sardar 2010, 183–184). Let us start with an obvious example for such misbalance: Nanotechnology.

Elitist Visions and Profane Desires

One of the emerging technologies that attracted most of TA's attention in the past 20 years is nanotechnology (Grunwald 2019, 53; Nordmann 2006). While the origins of the "idea" of nanotechnology are usually traced back to an after-dinner talk to the American Physical Society and subsequent publications by Richard Feynman (McCray 2013, 117), it is undisputed that the activities and publications of Eric Drexler – the "apostle of Nanotechnology" – triggered an avalanche of public debates, media responses, and a variety of research endeavors under the financial umbrella of the National Nanotechnology Initiative (Amato 1991; McCray 2013; Rip and Voß 2013). Even though there were a number of legitimate arguments to resist the vision's pressure to respond to it in certain ways (Nordmann 2007), the philosophical, political, and sociological reflections on the emergence of nanotechnology have gained a momentum, leading to new journals (*Nanoethics*), research paradigms ("vision assessment"), and insights about the development of new technologies. Nanotechnology in this manner is an exemplary technology to which meaning has been attributed before any concrete artifact existed. In the twilight zone of nanotechnologies material not-yet-existence and its making, the ubiquitous vision of nanotechnology, as advocated by Drexler, Roco and Bainbridge, and Bill Joy, has created a sense of urgency in TA, science and technology studies (STS), FS, and other fields to be concerned about the promises of a technology of which little was known (Drexler 1986; Joy 2000; Roco and Bainbridge 2003). The ascription of meaning induced a prescriptive power, the demand to respond to it in certain ways, for instance, by requiring justification from those who would have preferred ignoring nanotechnology's speculative facet altogether (Nordmann 2007).[2]

One of the central insights of this development from the previously advocated hermeneutic perspective, which is also affirmed by authors from other fields such as philosophy of technology, FS, and STS, is that Drexler's idea and his promotional activities had an impact for research policy and eventually technological development despite the far-fetchedness of nanotechnologies vision in the 1980s. The vision spurred Drexler and fellow scientists and functioned as medium or prism to negotiate the socio-technical future. Naturally, this development spawned an extensive debate about a meaningful notion of responsibility for highly contingent and far-reaching technological developments (Cabrera Trujillo 2014; Simakova and Coenen 2013; Sand 2016).

However, how exactly did nanotechnology develop the momentum it developed? How did its promises extend into popular culture and reached the realm of political decision-making and the attention of important stakeholders? In his book "The visioneers. How a group of elite scientists pursued space colonies, nanotechnologies, and a limitless future," Patrick McCray describes meticulously how Drexler and other so-called "visioneers" like Gerard O'Neill "[developed] a broad and comprehensive vision for how the future might be radically changed by technology, doing research and engineering to advance this vision, and promoting one's ideas to the public and policymakers in the hopes of generating attention and perhaps even realization. Throughout all these diverse activities, people like Drexler and O'Neill worked to build technical and social foundations for their own particular conceptions of the technological future" (McCray 2013, 13). McCray is eager to emphasize that advancing and realizing one's technological future requires financial support and suitable networks. He writes:

> "Like traditional engineering, visioneering requires money. Promoters of radical new technologies create a bricolage of patrons and supporters to finance their work. Many visioneers operated to some degree outside the patronage system that funds and supports conventional American scientists and engineers at universities, corporate labs, or federal facilities. Sometimes this was a deliberate choice, as it provided freedom without managerial oversight and Peer review. […] Venture capitalists, wealthy entrepreneurs, and curious citizens all contributed in varying degrees. To help raise funds, both men [Drexler and O'Neill] setup nonprofit institutes that helped promote their visioneering."
>
> (McCray 2013, 12)

There is clearly much to say (and much has been said) about the attractiveness and elasticity of nanotechnology as an umbrella term (Rip and Voß 2013) fueling our creativity and spurring imaginations of its application ranging from "nanobots" (Nerlich 2005) to visions of radical human enhancement through convergence of nano and other technologies (Coenen 2010; Roco and Bainbridge 2003). Furthermore, the semantic openness of such terms, their connectivity and elasticity, have been shown to enhance their potential for usage in public discourses and their fruitfulness for science journalism and popularization (Dickel and Schrape 2017). However, McCray's remarks indicate that it is not simply the content or design of a narrative that constitutes its appeal and suffices for its "success" (whether that being its realization or being considered by others as significant to a certain extent). The social backgrounds of visioneers, their country of origin, their education (technological expertise), their gender, and a variety of other features also affect the traction of their socio-technical visions and the activities surrounding them, maybe even more than the actual "nature" or guise of the visions themselves. I will subsume these aspects together in the following as the social standing of visionaries and suggest that this constitutes their power in a process, which has recently been termed "future making" (Sand and Schneider 2017; Schneider and Lösch 2018).[3] Although it has been mentioned before that power is a central aspect in these dynamics (Callaghan 2018; Grunwald 2012, 22; Sand and Schneider 2017), it is crucial to further clarify what this means and emphasize its significance from the point of view of hermeneutic TA. This is important also because it can be suggested that social standing as introduced before does not only affect whether one's vision in a discourse will or will not receive attention, but also how the future is perceived (open/closed) and, thus, whether one actually has a future.

To get to the heart of the issue, we should contrast the high-tech vision of nanotechnology with the rather "earthly" issues that concern a majority of people on the globe every day. We might realize that the pursuit of nanotechnology is neither at the fore of these peoples' interests nor within their sphere of comprehension for reasons that we shall explore. A remarkable anecdote in McCray's "Visioneers" can be found in the first chapter, which shall be briefly recalled in the following. In August 1977, space colonization advocate Gerard O'Neill, the author of "The High Frontier," an award-winning book on the subject of space colonization and founder of the Space Studies Institute, was invited to present his ideas to a thousand or so enthusiasts at the first so-called Space Day at the Museum of Science and Industry in downtown Los Angeles. While "Trekkies," NASA representatives, and other stakeholders and space enthusiasts gathered inside the museum, Robert Crumb, an artist and one of the few critical participants of the event, witnessed a small demonstration outside. McCray notes frivolously: "While Crumb fumed outside the museum, he could have encountered laid-off workers waving signs proclaiming 'Jobs on Earth, Not in Space' and 'Brown, Hire an Earthling' that expressed more terrestrial concerns" (McCray 2013, 3–4).

In this remarkable anecdote, we find a tantalizing confrontation between different visions of the future and subsequently a challenge for policy-making, when pursuing a common future: On the one hand, there is the far-fetched, high-technological vision of space colonization. On the other hand, there is what seems in contrast to be the somewhat "profane" desire for more job opportunities. Obviously, the wish for employment is anything but profane; employment is constitutive for peoples' well-being and happiness. Yet, in the light of the passionately promoted narrative of space colonization of the visioneer Gerard O'Neill, employment seems to be a concern of minor importance. As argued before, the envisioning of such technological pathway instantly encourages a typical set of questions regarding the feasibility and desirability of this future, questions with which TA and STS have been repeatedly concerned. It also leads to a redirection of focus – away from issues such as unemployment. The vision's advocacy indicates a value judgment suggesting that other problems such as unemployment can either be resolved by realization of the vision in the long run, as an incorporated but secondary value so to speak, or simply as being unimportant.[4] It should be emphasized that the previous observations are purely descriptive: Neither do I want to justify that an advanced vision of space colonization disregards employment as a value or central source of people's happiness, nor do I want to suggest that people who demonstrate for more job opportunities are simply unable to transcend their individual desires and regard their concerns from a more external point of view and in light of the needs of society or indeed mankind as a whole, which might be the viewpoint naturally adopted by advocates of space colonies. Instead, it should be pointed out how existing misbalances in power and social standing, which have so often been pointed out (Callaghan 2018), reappear in the ways in which people approach the future, what they envision, and whether they will be able to express these narratives in a manner that increases their momentum in the process of negotiating the future. From a reflective, hermeneutic perspective that directs our attention from the futures of emerging technologies to their present. Hence, we acquire an increased awareness of the relation between social standing and future making. It is obvious that these observations are not confined to the examples of Drexler and O'Neill and their visioneering activities. Ray Kurzweil, the former CEO of Engineering of *Google*, springs to mind, who founded several companies and the *Singularity University*, a hub for Artificial Intelligence (AI) enthusiast, and author of several best-selling books on the phenomenon of Singularity and AI (Joy 2000; Kurzweil 1999; McCray 2012). In a commentary to a recent Special Section in *Nanoethics*, Patrick McCray recapitulated his research on visioneering and carefully hints at a more general conclusion about the issue at hand:

> "I think the longer history of visioneering shows the clear advantage that technologists from elite backgrounds and privileged positions enjoy in developing their visions, finding resources to advance them, and getting the attention of the public and policy makers. I wouldn't go so far to say that visioneering is, *by nature* exclusive, but history suggests that it, *in practice*, often is."
>
> (McCray 2017, 206)

In the following two sections, I will analyze this more general claim and explore its normative implications.

Forestalling Inequalities in Future Making

There are a number of metaphors that connect the notion of time to certain situations or decisions, for instance, when we say that "someone's time is running out." This notion is ambiguous: It can mean that someone is forced to decide or fulfill a task within a finite amount of time, or, simply, that the person is about to die. Keeping the previous discussion and our specific field of application in mind, we should ask, what does "not having a future" mean in the context of the present debate?

When we say about a project or an endeavor that it does not have a future, we usually mean that it is doomed to failure. This does not seem to apply to the demonstrators in L.A., which have been portrayed before. They are not doomed to fail. Instead, they might be the first representatives of a very influential social movement (Note that I am here primarily interested in analytic distinctions, not historical accuracy). In the final chapter of her recent thought-provoking book "The Ethics of Invention," Sheila Jasanoff suggests that "not having a future" means lacking the anticipatory capacities to assess the long-term prospects of emerging technologies. She also suggests in vein with my previous argumentation that the misbalance in future making constitutes an ethical problem, just like social and political inequality:

> "[…] most of the world's masses are in no position to anticipate for themselves either immediate benefits or improved long-term prospects from the forward march of technology. They must accept the promise of benevolent outsiders that their lives will be bettered through inventions designed elsewhere, by entrepreneurs closer to technology's moving frontiers, with the capital and know-how to engineer large-scale change. Inequality–not only as access but even more of anticipation–thus emerges as an unresolved ethical and political barrier to the just governance of technological innovation."
> (Jasanoff 2016, 255–256)

This paragraph neatly summarizes the heart of our previous analysis. In a central regard, however, it also goes beyond the previous proposal by highlighting a specific aspect. Many people are passive bystanders of the techno-visionary dynamics initiated by rather exclusive communities of tech-affine people in the Northern Hemisphere. Beyond being simply powerless in making a change to certain visionary discourses, Jasanoff suggests in this passage that many people also do not entertain their own narrative of what desirable change could mean. They are unable to establish their own (socio-technical) prospect, which could potentially compete in the arena of futures with proposals by more powerful and exclusive tech visionaries. Jasanoff suggests that a lack of voice is not the problem: Rather, the problem seems to be that many people have nothing to say. This emphasizes a somewhat different connotation than the one discussed before and puts us in a great position to distinguish and categorize the different notions of "not having a future." As this task is initially explorative, I will refrain from suggesting that any one of these or a group of them has a prerogative and captures the meaning of "not having a future" more accurately than the others. Note also that this list is neither comprehensive nor are these notions exclusive: Two or more of these notions might apply to a person or a group at the same time. The following conceptual distinctions might form the backdrop for empirical studies of various political or social groups and their attitudes and stances towards social or technological futures.

I People, who do not have a future, have a short life expectancy. They will not live long. This meaning of "not having a future" is quite literal and has a semantic overlap with our interpretation of "someone's time is running out." It is fair to assume that this is not what our concerns are centered around when discussing techno-visionary discourses and dynamics. We shall, therefore, not discuss this notion any further.

II People, who do not have a future, lack anticipatory capacities: They lack education in certain disciplines (physics, economics, information technologies, biology, etc.) making them ignorant regarding the state of the art of the science and the research landscape of certain techno-sciences which undermines their ability to make suggestions about their possible futures (Jasanoff).

Jasanoff emphasizes this meaning at the end of her book. Having identified this as a problem, a large variety of methods ranging from scenario planning and anticipatory action research to science-fiction writing have been proposed by TA and FS to encourage people to create their own visions of the future despite a lack of expertise in specific fields (Miller and Bennett 2008; Stevenson 2002). This is suggested to alleviate the epistemic blind spots in planning processes by bringing unthought-of possibilities to the fore of decision-makers, multiplying the sources of creativity and, thus, enabling to choose from a wider range of pathways and prepare for a wider range of possibilities, making political decisions more *robust* and legitimate (Gudowsky et al. 2012; Grunwald 2019, 64; Ramírez and Selin 2014). Eventually, this approach forestalls the social construction and co-production of technologies (Bijker et al. 1987) with the social construction of futures (Fuller and Loogma 2009). In general, education is often considered as a means to enhance people's capacity to envision the future of a certain techno-science; education seems to be one constituent for successful anticipation (Railton 2016, 6).

III People, who have lost *hope*, might be considered as not having a future. They lack the firm conviction that their present desires will ever be fulfilled in the future. This attitude can – but does not necessarily – imply a refusal to anticipate one's own or societies' future beyond the most existential, momentary desires for food and shelter, which might be relinquished only in cases of severe despair or mental illness (depression).

We can imagine, for instance, people who received a life sentence in prison. Consider, for instance, the protagonists of Frank Darabont's "Shawshank Redemption." After spending years in this facility and being repeatedly exposed to acts of humiliation, a person, who serves a life sentence in prison, might still *dream* of a better future, of a life outside the prison, of having a family and kids, and so on. However, he has given up *hope* that such future will ever materialize. Here, we should not dwell on the question whether it is reasonable to be rather hopeful or despaired when the likelihood of a certain event such as being released from prison is rather low (Meirav 2009, 223). I also want to neglect at this point the subtle differences between the concepts of "optimism" and "hope" because of their irrelevance for the present discussion. What is more important firstly, is that hope can affect one's attitude in terms of trust in the occurrence of future events and this attitude can be affected by *social standing*. It is obvious, for instance, that people who are victims of oppression or social exclusion can be systematically hindered to act on their values or desires. Consider a homosexual person, who lives under a conservative, religious regime, or a journalist who lives in a country in which freedom of press is eroded. It is understandable that such people doubt their life's meaningfulness (Oakley 2010, 113–114). The oppression, however, does not have to be as severe as in these exemplary cases to undermine optimism: A few years of joblessness can take away any hope for future employment – a staggering prospect. Secondly, it must be mentioned that while optimism and hope regarding the future might be affected by social standing, there is no conceptual (analytic) connection between hopelessness and a lack of vision or anticipatory capacity. As mentioned before, it is not unreasonable to entertain a vision despite its improbability: Even a lifelong prisoner can daydream of a life in freedom. If she is hopeless, she gave up believing that this prospect will ever materialize. On the other hand, one can be extraordinarily hopeful regarding the future, without having a concise idea (a vision) of what might come. The optimistic slogan "Everything is going to be alright" is *semantically open* and applies to all kinds of future realities.

IV People, who lack a comprehensive (technological) vision of a future society commensurable with the idea of technological singularity, space colonies, nanotechnology, or human enhancement, do not have a future.

We remember the previously sketched confrontation between the advanced, high-tech vision of space colonization by O'Neill and what has been termed for lack of a better concept the "profane" desire of employment (which is, to repeat, anything but profane). In this sense, people might have a vision of the future at whose center we find social reform instead of technological innovation. The particular parameter that is envisioned – such as employment in our example – links the individual desire to a more general social state. Here again, the desire and vision, even the additional motivation to demonstrate and participate in a political movement (or party), might be detached from being actually optimistic or hopeful about the prospect of such sociopolitical change. While one might be tempted to consider such people as "not having a future," caution has to be dispensed before such judgment is made. To begin, we need to distinguish clearly between people who simply desire a job for themselves, without much of an idea how employment in their neighborhood can flourish and those who might not even be unemployed but understand the importance of the issue for society and develop a political agenda to challenge them. The former is a more basic notion of envisioning a future with a stronger focus on one's individual desires (this notion has close proximity to the second one described above). It attributes the responsibility to deliver policy plans to increase job opportunities to leading political bodies and social ministries.[5] The second envisions a more comprehensive picture and suggests possible ways to realize this future. Such person might become a spokesperson of a union or a more or less structured interest group or movement (Brettschneider and Schuster 2013). Again, emphasizing this distinction here serves purely analytic purposes.

Although the second type seems to be more profound, both of these types are central in a liberal democracy in which not everyone is obliged to become a politician or an expert in other fields that concern their lives. I might desire a healthy life and still leave the responsibility to ensure that this materializes in periods of illness to my doctors. Modern states are functionally highly "differentiated" in distributing this sort of real and "cognitive labor" (Kitcher 1990), which is an important point to which I shall come back later on (Luhmann 1977). As both imagined people *have a future* in the just described manner, we should reconsider the previously made judgment.

Most strikingly in the first case, but also in the second, these futures do not seem to be *commensurable* to the visions usually discussed in TA such as nanotechnology, synthetic biology, in-vitro meat, and many others (Grunwald 2014b). Firstly, this seems to be because at least one of our exemplary demonstrators might have an individual desire for employment but lacks a *comprehensive* vision of the future such as those envisioned and promoted by Eric Drexler in the "Engines of Creation," or Kurzweil's "The Age of Spiritual Machines." Secondly, they both lack a direct connection to technological innovations, which are at the center of TA research, as its name already suggests. Both of these suggestions are flawed. A sufficient amount of people who desire employment can form a movement that easily threatens even large-scaled technological endeavors (Larsen 2003): Consider the protests against "Stuttgart 21," a railway-station project in the capital of the German federal state Baden-Württemberg (Brettschneider and Schuster 2013; Grunwald 2019, 70). The project aims at rebuilding the railway terminus Stuttgart to become a thorough railway station. From the beginning, it was criticized as unnecessary, expensive, and stressful for local residents and the environment. The protests, in which up to 100.000 people participated, cumulated in a violent confrontation between protesters and the police on September 30th, 2010, where several hundred people were injured. Subsequently, the reigning conservative federal government was replaced by the Greens, who advocated an anti- "Stuttgart 21" policy and a referendum was held. The 300 trees in the castle ground that had to be chopped for the building project became an epitome of the destructive consequences of "Stuttgart 21" (Novy and Peters 2012, 134).

While such futures – "trees before railway station" – might be "narrow" and ignorant of technological possibilities, they can have a tremendous impact on technological projects. This applies even more to the second type of imaginary demonstrator, who might become an extremely knowledgeable layperson in the field of labor market policy: Although "Stuttgart 21" is still pursued, a number of lasting collectives and parties emerged from the anti-"Stuttgart 21" movement, such as *Stuttgart Ökologisch Sozial*, which continue to be politically active in the local municipality and which contributed at the time as self-educated experts to establishing a more satisfactory station model. Therefore, while there might be an incommensurability in the *form* of those futures (comprehensive/narrow, narratively outlined/condensed in a buzzword or slogan, technological/non-technological), the confrontation at the Science Museum in Los Angeles sketched above is not between people who have and others who do not have a future. The confrontation rests on a divergence in value ranking: Some futures emphasize societal change or environmental matters as the primary concern and rank technological projects as secondary. O'Neill's future displayed an inverse ranking: Technology first. Hence, while there is a sense in which the demonstrators outside the museum "have no future," their futures are highly relevant for TA research. Thinking that technological futures could be challenged, redesigned, or undermined only by narratives of similar texture, form, or content is flawed. There is one more meaning of "not having a future" to consider, before I will draw some more general conclusions.

V People lack the means to make their visions gain momentum in societal discourses: They do not have a voice in policy-making.

When we say that someone does not have a future, we can also mean that people's futures do not matter in decision-making processes. They do have a future (with or without technology at their centers), but these futures are either ignored or dismissed when plans are established and projects are being launched. Here too, we can refer to the situation at the Science Museum in Los Angeles in 1977. The speakers of the Space Colonization movement presented their views on stage inside the museum, while important decision-makers were invited to join the discussions and be persuaded by the presentations (McCray 2013). Because of his academic affiliation and involvement in expert communities, Gerard O'Neill had contacts to powerful stakeholders and decision-makers. Outside the museum, there were no stages, drinks, or invited guests. This might just indicate whose voices will be heard. FS researcher Tony Stevenson notes on this account:

> "Creating the future can be controlled by the wealthy, powerful and famous, and their minders and lackeys. But in the spirit of democracy, future-creating would seek to ensure that people who have a stake in the future, either through their likely habitat there, or their successor generations, should be able to participate in that creation. This does not happen with the more traditional methodologies of futures studies, where experts stand aside from the vast majority of other citizens."
>
> (Stevenson 2002, 418–419)

In this passage from his "Anticipatory Action Learning: Conversations about the Future," Stevenson seems to equalize the encouragement towards creating futures with their control, emphasizing, however, that control often lies in the hand of the powerful and wealthy. It is very important to distinguish these two notions, something that we have neglected so far. For several years, TA, FS, and STS encouraged the envisioning of other pathways by employing scenario methods and other techniques that aim at increasing the number of possible pathways, diversifying the future, and thereby challenge epistemic blind spots and enhance creativity ((Ramírez and Selin 2014; Sardar 2010, 183). Jasanoff's proposal, although it is not explicit in this regard, can be understood as a demand to counteract the widespread issue of "not having a future" with supporting the advancement of anticipatory capacities and encouraging future making by involving people into scenario building and anticipatory action methods to create ever more futures. While such development is in most cases praiseworthy, it must be underlined that this cannot ensure that these futures ever materialize. There are two obvious obstacles to this: Regarding the making of the future, the previously mentioned institutions' (TA, STS, and FS) power is limited.

Even if these institutions were globally active and enhanced anticipatory capacities worldwide (which is a mammoth task and obviously far from being a reality), they are only one amongst a host of societal agents that together (not equally) determine which future is pursued, and certainly not the most powerful either. Secondly, the creation of more futures does not erase the difficulty of choosing between them, it might be suggested that it rather increases this challenge. The more futures are on the table, the smaller the probabilities for each to materialize. If we think of this like an ideal card game – which, of course, it is not – the likelihood of picking a specific card gets smaller, the higher the stack gets.

Five different meanings of the notion "not having a future" have been distinguished and discussed. Between the lines, I have insinuated some (normative) pitfalls, which should be outlined in more detail in the following section.

Conclusions and Outlook – Pitfalls of Democratizing the Future

Hermeneutic TA is understood and advocated as a critical perspective on the development and distribution of futures and visions *in the present* because of the typical epistemic problems underlying predicting the future. Hermeneutic TA aims at a better understanding of how and why some futures gain traction in current discourses and why and how they are attributed with meaning. Throwing a reflective and critical light on these dynamics also means, as I have argued introductorily, considering TA as a player that does not only function as a neutral observer, but also as a significant participator in this process. Because of this, I encouraged to explore TA's focus on some particular objects of study like nanotechnology instead of other (non-technological) visions about the future. I suggested that this truly hermeneutic stance might enhance our understanding of envisioning and anticipating as social activities and reveal knowledge about the present. From this standpoint, it becomes clear that TA has focused on high-tech visions of few elitist, powerful academics from the Northern Hemisphere, and neglected rather "profane" futures. Given this rather narrow scope on future making, we might be inclined to say of many people that they do not have a future that will ever come to the fore of policy-makers, stakeholders, FS, TA, or similar such institutions. We have then explored the semantic landscape of this notion and distinguished five different meanings of "not having a future." More generally, we can understand the previous reflections as an emphasis of the issue of power imbalances in how visions are being produced and distributed.

It is important to note again that the presented account was intended to be primarily descriptive. There might be good reasons to neglect the magnitude of uncountable futures out there. There are certainly *prudential reasons* for such ignorance, as this endeavor is beyond the capacities of any existing institution. However, I also suggested that it is not always the comprehensive, technological narratives that shape the future: If a sufficient number of individuals repudiate a certain technological future or even foster contradictory desires, the high-tech futures of white elites will not gain traction. This can happen, even if these individuals do not entertain their own alternative vision of a desirable future. For this reason, it would be equally wrong to neglect profane futures altogether.

As an institution with a social function, TA has continuously emphasized the importance of including citizens and stakeholders in future making (Grunwald 2019, 63 ff.). Behind many such projects stands a democratic ideal, but also epistemic considerations (bringing previously unthought-of possibilities to the table) and practical values (increasing the level of acceptance of new technologies, stabilizing them, and, thereby, making them more sustainable and resilient). In this vein, one might suggest that it is indeed pitiful that many people do not have a future. One might be inclined to defend an obligation or an imperative for the ubiquitous advancement of anticipatory capacities to provide everybody with the opportunity to participate in the making of the future and thereby challenge the prevailing inequalities outlined above. This is a noble and praiseworthy impulse and, in most cases, it means betterment. As an outlook, I will nevertheless attempt to outline some of the pitfalls of this motivation that shone through the lines of the previous argumentation:

i As mentioned before, we must resist considering comprehensive, socio-technical visions such as nanotechnology, which have long been under scrutiny for contingent reasons of access and power, as the only basis of policy-making and the only adequate basis for scenario building. In contrast, we should learn to acknowledge that "profane" visions lacking technological content (like the desire for employment) are anything but negligible, just because they are incommensurable with high-tech visions. It seems that this failure gave rise to the initial assumption that many people do not have a future. Factually, such "profane" desires might well become the starting point of a technological or an anti-technological movement that alters societies more profoundly than any currently emerging technologies (again, think about the social repercussions of the anti-"Stuttgart 21" movement). We should start to think differently about such "narrow" narratives and maybe learn to acknowledge the wish for employment as a substantial vision that implies a value ranking, downgrading technological progress to second or an even lower position. Hence, the impulse to enhance these people's anticipatory capacities imposes an alien understanding of envisioning the future on them that demands a reassessment of their desires in light of something they might not actually value, namely a high-tech future.

ii There are certainly many people, who want to participate in discourses about the future, who lack the anticipatory capacities to establish their own visions and launch them into the discourse, and who would, therefore, highly appreciate partaking in educational programs to improve their anticipatory skills. This, however, does not apply to everyone. "Not having a future" can be a choice indeed. One can choose to brush the responsibility for the envisioning and making of the future aside and trust that the futures of experts from different fields yield the best outcome for oneself. Consider, for example, that it is not unusual that people who save money and to receive interest rates, consign it to banks or other financial institutions and let them handle it. One assumes then that banks' investment plans will create more revenue than oneself possibly could. Such people's future is based on a trust into the visions of *these institutions* regarding their ways of increasing people's interest rates and a delegation of responsibility for complex and momentous decision, which they themselves do not want to bear. Such attitude is not *a priori* blameworthy. Our impulse to enhance anticipatory capacities must acknowledge that modern, liberal societies are constituted by a division of labor allowing people to choose to be guided and governed by other people of whom they believe to produce the more reliable and resilient narratives of the future for them and society in general. Although, it might be a difficult lesson to learn: It is everyone's right to "not have a future" and to trust the experts of a certain field. In practice, this means giving people a right to refuse participating in anticipatory actions and scenario building workshops.

iii Even if many people participate in anticipatory actions and scenario building methods and subsequently develop their own visions, we should not naively assume that this indeed leads to a more democratic future in which more desires are fulfilled and more visions realized. Power imbalances appear not only in the envisioning of the future, but also in their realization. Even if we manage to advance people's capacities to anticipate and create more futures, choosing some of this growing stack of possible pathways for realization remains a difficult and open task. Clearly, policy-making is not like chess playing, where one's allowed to make only one distinct move per round and where it is unequivocally determined what counts as a legitimate move. Policy-making can find compromises between diverging views or support the renunciation or amelioration of some views in light of competing interests.

Nevertheless, the practical difficulties to realize everyone's visions undeniably persist as various visions likely stand not only in a complementary but in mutually exclusive opposition to each other: How then should one decide which ones to pursue? This issue does not disappear with the advancement of anticipatory capacities: It might become even more frustrating for those who finally envisioned an own future, when they realize that it will not ever materialize.

Notes

1 This article was first published in a slightly revised version in *Futures*: Sand, M. 2019. "On 'not having a future'." *Futures* 107: 98–106. doi: https://doi.org/10.1016/j.futures.2019.01.002.

2 Synthetic biology is clearly another field where imbalances in future making have occurred and pertain: In a report by the ETC Group, a Canadian CSO, on "Extreme Genetic Engineering-An Introduction Into Synthetic Biology" we can read about an exclusive "Asilomar-like" meeting of leading scientists in synthetic biology in Berkeley in 2006, where "no civil society representatives

were in attendance – those who tried to register were turned away due to 'limited space'" (ETC Group 2007, 47). This setting resembles in various ways the exclusive Space Day meeting from 1977 described below. The ETC report expresses numerous concerns about questions related to power and control and "a lack of societal debate about synthetic biology" (ETC Group 2007, 4). These concerns were picked up in deliverables and transformed into participatory actions by TA institutions such as the *Rathenau Institut* (Stemerding et al. 2013, 16).

3 I am well aware that this notion is utterly imprecise: It is of course not the future that is made, when new technologies are developed or other techno-scientific research projects are undertaken. It is this research or these technologies that are being made. "Making the future" here refers to a host of such projects.

4 It has been shown that socio-technical endeavors that try to make believe "that nothing special is undertaken, which considers its dreams of control and improvement to require little external endorsement or explanation, and which is embedded within a set of master narratives in which science and technology are staged unambiguously as the solution to a range of social ills," reinforce pessimistic and bleak views in the general public about the prospects of nanotechnology (Macnaghten 2010, 32–33).

5 A 2015 report of the Office of Technology Assessment at the German Bundestag (TAB) notes that "the general public is hardly aware of the scientific development [of synthetic biology] at this early stage and thus would have to be **motivated actively** to deal with the issue [own emphasis]" (Sauter et al. 2015, 16). This does not mean that the public does not entertain an idea about the future of synthetic biology: This future could be that someone else – a group of policymakers or experts – should act in societies' best interests and ensure its safety and desirability

References

Amato, I. 1991. "The Apostle of Nanotechnology." *Science* 254 (5036): 1310–1311. 10.1126/science.254.5036.1310.

Bijker, W. E., T. P. Hughes, and T. J. Pinch. 1987. *The Social Construction of Technological Systems: New Directions in the Sociology and History of Technology*. Cambridge, Mass: MIT Press.

Brettschneider, F., and W. Schuster, eds. 2013. *Stuttgart 21- Ein Großprojekt zwischen Protest und Akzeptanz*. Wiesbaden: VS Verlag für Sozialwissenschaften.

Cabrera Trujillo, L. Y. 2014. "Visioneering and the Role of Active Engagement and Assessment." *Nanoethics* 8 (2): 201–206. 10.1007/s11569-014-0199-5.

Callaghan, C. W. 2018. "Surviving a Technological Future: Technological Proliferation and Modes of Discovery." *Futures* 104: 100–116. 10.1016/j.futures.2018.08.001.

Coenen, C. 2010. "Deliberating Visions: The Case of Human Enhancement in the Discourse on Nanotechnology and Convergence." In *Governing Future Technologies: Nanotechnology and the Rise of an Assessment Regime*, edited by M. Kaiser, 73–88. Sociology of the Sciences Yearbook 27. Dordrecht: Springer.

Dickel, S., and J.-F. Schrape. 2017. "The Logic of Digital Utopianism." *Nanoethics* 11 (1): 47–58. 10.1007/s11569-017-0285-6.

Drexler, E. K. 1986. *Engines of Creation: The Coming Era of Nanotechnology*. New York: Anchor Books.

ETC Group. 2007. *Extreme Genetic Engineering-An Introduction to Synthetic Biology*. Ottawa: ETC Group Report.

Fuller, T., and K. Loogma. 2009. "Constructing Futures: A Social Constructionist Perspective on Foresight Methodology." *Futures* 41 (2): 71–79. 10.1016/j.futures.2008.07.039.

Grunwald, A. 2012. *Technikzukünfte als Medium von Zukunftsdebatten und Technikgestaltung*. Karlsruher Studien Technik und Kultur 6. Karlsruhe, Hannover: KIT Scientific Publishing.

Grunwald, A. 2014a. "Modes of Orientation Provided by Futures Studies: Making Sense of Diversity and Divergence." *European Journal of Futures Research* 2 (1). 10.1007/s40309-013-0030-5.

Grunwald, A. 2014b. "The Hermeneutic Side of Responsible Research and Innovation." *Journal of Responsible Innovation* 1 (3): 274–291. 10.1080/23299460.2014.968437.

Grunwald, A. 2015. "Die hermeneutische Erweiterung der Technikfolgenabschätzung." *Technikfolgenabschätzung - Theorie und Praxis* 24 (2): 65–69. Accessed May 04, 2018.

Grunwald, A. 2017. "Assigning Meaning to NEST by Technology Futures: Extended Responsibility of Technology Assessment in RRI." *Journal of Responsible Innovation* 4 (2): 100–117. 10.1080/23299460.2017.1360719.

Grunwald, A. 2019. *Technology Assessment in Practice and Theory*. New York: Routledge.

Gudowsky, N., W. Peissl, M. Sotoudeh, and U. Bechtold. 2012. "Forward-Looking Activities: Incorporating Citizens' Visions: A Critical Analysis of the CIVISTI Method." *Poiesis and Praxis* 9 (1-2): 101–123. 10.1007/s10202-012-0121-6.

Jasanoff, S. 2016. *The Ethics of Invention: Technology and the Human Future*. The Norton Global Ethics Series. New York, London: W.W. Norton & Company.

Joy, B. 2000. "Why the Future Does Not Need Us." *Wired Magazine* 8 (4). http://www.wired.com/wired/archive/8.04/joy.html. Accessed September 18, 2018.

Kitcher, P. 1990. "The Division of Cognitive Labor." *Journal of Philosophy* 87 (1): 5–22. 10.2307/2026796.

Kurzweil, R. 1999. *The Age of Spiritual Machines: When Computers Exceed Human Intelligence*. New York: Penguin.

Larsen, K. S. 2003. "Oppression and Resistance: The Powerless and the Future." *Futures* 35 (2): 163–167. 10.1016/S0016-3287(02)00025-3.

Luhmann, N. 1977. "Differentiation of Society." *The Canadian Journal of Sociology* 2 (1): 29–53. 10.2307/3340510.

Macnaghten, P. 2010. "Researching Technoscientific Concerns in the Making: Narrative Structures, Public Responses, and Emerging Nanotechnologies." *Environment and Planning A: Economy and Space* 42 (1): 23–37. 10.1068/a41349.

McCray, P. 2012. "California Dreamin': Visioneering the Technological Future." In *Where Minds and Matters Meet: Technology in California and the West*, edited by V. Janssen, 347–378. Western Histories 6. Berkeley: University of California Press.

McCray, P. 2013. *The Visioneers: How a Group of Elite Scientists Pursued Space Colonies, Nanotechnologies, and a Limitless Future*. Princeton: Princeton University Press.

McCray, P. 2017. "Futures Perfect and Visioneering: A Re-Assessment." *Nanoethics* 11 (2): 203–207. 10.1007/s11569-017-0303-8.

Meirav, A. 2009. "The Nature of Hope." *Ratio* 22 (2): 216–233. 10.1111/j.1467-9329.2009.00427.x.

Miller, C. A., and I. Bennett. 2008. "Thinking Longer Term About Technology: Is There Value in Science Fiction-Inspired Approaches to Constructing Futures?" *Science and Public Policy* 35 (8): 597–606. 10.3152/030234208X370666.

Nerlich, B. 2005. "From Nautilus to Nanobo(a)ts: The Visual Construction of Nanoscience." *AZojono-Journal of Nanotechnology Online* (1): 1–19. https://www.azonano.com/article.aspx?ArticleID=1466. Accessed September 18, 2018.

Nordmann, A., ed. 2006. *Nanotechnologien im Kontext: Philosophische, ethische und gesellschaftliche Perspektiven*. Berlin: Akademische Verlagsgesellschaft.

Nordmann, A. 2007. "If and Then: A Critique of Speculative Nanoethics." *Nanoethics* 1 (1): 31–46. 10.1007/s11569-007-0007-6.

Novy J., and D. Peters. 2012. Railway Station Mega-Projects as Public Controversies: The Case of Stuttgart 21. *Built Environment* 38 (1): 128–145. 10.2148/benv.38.1.128.

Oakley, T. 2010. "The Issue is Meaningless." *The Monist* 93 (1): 106–122. 10.5840/monist20109317.

Ramirez R., and C. Selin. 2014. "Plausibility and Probability in Scenario Planning." *Foresight*, 16 (1): 54–74. 10.1108/FS-08-2012-0061.

Railton, P. 2016. "Introduction." In *Homo Prospectus*, edited by M. E. P. Seligman, P. Railton, R. F. Baumeister, and C. Sripada, 3–31. Oxford: Oxford University Press.

Rip, A., and J.-P. Voß. 2013. "Umbrella Terms as Mediators in the Governance of Emerging Science and Technology." *Science, Technology & Innovation Studies* 9 (2): 39–59.

Roco, M. C., and W. S. Bainbridge., eds. 2003. *Converging Technologies for Improving Human Performance: Nanotechnology, Biotechnology, Information Technology and Cognitive Science*. Dordrecht, Boston, Mass: Kluwer Academic Publishers. Accessed September 18, 2018. http://www.wtec.org/ConvergingTechnologies/Report/NBIC_report.pdf.

Sand, M. 2016. "Visioneering and Responsibility- Opening Pandora's Box." *Nanoethics* 10 (1): 75–86, 10.1007/s11569-016-0252-7.

Sand, M., and C. Schneider. 2017. "Visioneering Socio-Technical Innovations." *Nanoethics* 11 (1): 19–29. 10.1007/s11569-017-0293-6.

Sardar, Z. 2010. "The Namesake: Futures; Futures Studies; Futurology; Futuristic; Foresight—What's in a Name?" *Futures* 42 (3): 177–184. 10.1016/j.futures.2009.11.001.

Sauter, A., S. Albrecht, D. van Doren, H. König, T. Reiß, and R. Trojok. 2015. *Synthetic Biology – The Next Phase of Biotechnology and Genetic Engineering*. Berlin: Office of Technology Assessment at the German Bundestag, 2015: Working report no. 164.

Schneider, C., and A. Lösch. 2018. "Visions in Assemblages: Future-Making and Governance in FabLabs." *Futures* 109. Online First. 10.1016/j.futures.2018.08.003.

Simakova, E., and C. Coenen. 2013. "Visions, Hype, and Expectations: A Place for Responsibility." In *Responsible Innovation: Managing the Responsible Emergence of Science and Innovation in Society*, edited by R. Owen, J. R. Bessant, and M. Heintz, 241–266. Chichester, UK: John Wiley & Sons.

Stemerding, D., and V. Rerimassie. 2013. *Discourses on Synthetic Biology in Europe*. Deliverable. The Hague: Rathenau Institut: Working Paper 1305.

Stevenson, T. 2002. "Anticipatory Action Learning: Conversations About the Future." *Futures* 34 (5): 417–425. 10.1016/S0016-3287(01)00068-4.

7 On Profane Futures and Profane Futures Literacy

Christopher Groves

Introduction

In debates over the relationship between scientific expertise and "lay" perspectives on science, a persistent view is that "the public" entertains suspicions about scientists and refuses assent to some scientific ideas (such as anthropogenic global warming) or technological proposals (like genetically engineered organisms) mainly because of a lack of scientific understanding This deficit model of the public understanding of science has been the object of much criticism from within the sociology of science and science and technology studies (STS) (Sturgis and Allum 2004). Martin Sand (2019) has recently drawn attention to how a comparable deficit model is at work within "imagineering" or visioneering, and perhaps within anticipation studies itself, noting that there is often a distinction made within the work of visioning futures between what might be called elite or expert future visions on the one hand, and "profane" futures on the other.

Sand introduces the concept of profane desires (like the desire for continuing secure employment) and profane futures (which are visions which do not feature technological content and ideas about large-scale societal transformations that are dependent on technological innovation). Profane is a word that registers a perceived lack, though the lack here is not perceived by Sand himself, but by proponents of technological visioning, exemplified in Sand's article by Patrick McCray (McCray 2013). "Profane" here retains the double meaning of being both uninitiated into esoteric knowledge and being the opposite of that which may be considered to be technological late modernity's characteristic form of the sacred, that is, technological visions themselves. Sublime technological visions may be seen as an extension of a transformation begun in the 19th century in which the idea of progress becomes the "mechanization" (*Technisierung*) of the future (Hölscher 1999). In 18th century Europe and America, revolutions had sought to refound political communities, not around the sanctity of either the historical or immemorial past, but instead around a consensus built on a new civic future built on equality before the law. The 19th century, by contrast, saw the scientization of emerging European and north

DOI: 10.4324/9781003322290-9

American social states. In this phase, the future-oriented ideal of formal equality was supplemented by ideas relating to the social provision of needs, focusing on the improvement of infrastructure and then of health, education and so on. The use of natural and emerging social sciences as means for increasing the efficiency with which needs could be met became enshrined within governance as the century drew to a close.

The Arendtian irony of the mechanization of the future is, therefore, that the 18th-century European struggles over republican ideas issued in an emerging focus on the profane, on the more effective administration of the social through ever more perfect *techne*. The managerial administration of needs, which may in one way be a more inclusive and wider ranging focus for futuring than visions which concern themselves with political rights, may however turn out to be exclusionary. The work of creating visions, as recent work in STS emphasizes (Konrad et al. 2016), is a central form of labor through which actors in post-traditional societies contribute to both the reproduction and transformation of the societies they inhabit. As Sand points out, this centrality of visioning to the administration of modernity means that some people are then held to suffer from a visioneering deficit. They do "not have a future" in the sense that they "neither entertain their own vision of a desirable socio-technical future [...] nor do they have the means to strategically position them in the discourse and, thereby, contribute to their realization" (Sand 2019, 99).

Sand discusses four ways in which the meaning of this lack of a future could be more precisely understood, which I paraphrase here.

1 People may lack anticipatory capabilities, enabling them to reflect on a wider range of future socio-technical possibilities. This implies a lack of "futures literacy" and "knowledge of how to use the future; it is familiarity with anticipatory processes" (Poli 2015, p. 110). Futures literacy implies being comfortable with ways of formally expressing future possibilities, like risk, and also with ways of responding to deeper uncertainties, e.g. with contingency planning. More fundamentally, it implies an acknowledgment that representations of the future concern possibilities that are, at best, partial expressions of present latent tendencies. Further, it implies an awareness that representation of the future is always in intervention in these tendencies out of which future states will emerge.
2 People may lack hope in the future (they may dream about one, but do not expect to realize it). This kind of condition has been identified by Arjun Appadurai as an erosion of a specific capability, or rather metacapability, that of aspiration, and thus an injustice (Appadurai 2013), a loss not of any particular envisioned future, but of belief in the future as such as potentially better.
3 People may not focus on technological visions of transformation, but they may have other comprehensive visions of the future that draw

on other sources of expertise than those which underlie technological visions.

4 People lack the means to ensure that their visions gain traction in societal discourses.

If we accept that these ways of "lacking a future", ofbeing on the wrong side of the profane/sacred distinction, are reasonably accurate, how should we understand the processes out of which these inequalities between actors arise? I want to suggest that the "lack" here is not so much a deficit on the part of those who "do not have a future," but instead stems from a failure of recognition that can be considered unjust and oppressive. This failure of recognitionalso helps to produce particular dominant knowledge practices as reliable means of constructing actionable futures for and by other actors, and in this way helps to create and maintain conduits of power.

Beyond Deficits

STS developed early on understandings of the relationship between lay perspectives and expert ones that represented alternatives to the deficit model. For example, lay perspectives were seen as potentially contributing to a "sub-politics" of expertise in which scientific knowledge, alerted through activism about the potential unintended consequences of technological innovation, could be turned reflexively against itself in order to question whether scientific research was necessarily aligned withbroader societal priorities (Beck 1992). Alternatively, lay perspectives could be leveraged as alternative forms of expertise in the social assessment of the implicit priorities or biases of scientific research and technoscientific mobilization (Funtowicz and Ravetz 1990; Wynne 1996).

The traditional public understanding of science deficit model fails to recognize the wider relevance of public views on the priorities and practices of technoscience, and thus can create situations of epistemic injustice (Fricker 2007) in which the credibility of relevant knowledge possessed by individuals is systematically denied. What I want to suggest is that the view that some futures are "only" profane represents a similar injustice, and like epistemic injustice more widely, is one which is created by specific social conditions. Of the four meanings of a "lack of future" Sand identifies, (3) and (4) clearly represent forms of epistemic injustice (recognition and procedural injustice, respectively) similar to those created by the deficit model of public understanding. Insofar as it represents a loss of capability of the kind Appadurai describes, (2) is also a clear case of distributive injustice conditional on (3) and (4). By contrast, (1) is an ambiguous description, for reasons that we now turn to explore, and which, I would suggest, underlie particular instances of (3) and (4).

In one sense, no living being is without a future, so long as we understand living organisms as anticipatory systems (Poli 2010). Having a future is, at one level, inextricably part of all human experience. Phenomenological traditions in philosophy and sociology represent the anticipatory projection of potential futures projected beyond the present through embodied emotion, imagination, and conceptual thought as essential for the organization of belief and action, and for meaning-making more generally. Futurity, along with remembrance, in this sense is a meaning-conferring horizon of consciousness that supervenes upon bodily rhythms and habits, and is thus part of more fundamental modes of anticipation understood as a metacapability of a living system. As Niklas Luhmann puts it, "the essential characteristic of a horizon is that we can never touch it, never get at it, never surpass it, but that in spite of that, it contributes to the definition of the situation" (Luhmann 1976, 140). The nature of this horizon bears further exploration, however. As Appadurai's ideas about a "loss of future" testifies, the size and reach of this horizon can change.

The future, as a dimension of experience, is for human beings primarily lived through relationships – with significant others, places, institutions, and ideals which can be considered objects of emotional attachment that support both individual identities and sometimes collective ones (Groves 2014; Honig 2012). This lived future (Adam and Groves 2007) is one worked on through the construction of narratives in which attachments provide structure, connecting the persisting past through a passing present to an insisting future. So long as people possess attachments, they are implicated in this concern and involvement with the future. Given that this involvement is dependent on a "convoy" of diverse attachments (Antonucci and Akiyama 1994), the future as projected and narrated by individuals beyond the present always has a particular character. Not only does the future as horizon project more or less further beyond the present, it projects more broadly or more narrowly. Given that this future is tied to the fate of attachments, its still latent potential is always affectively and emotionally colored. A lived future can, for example, be too much to bear, desperately wished for, too troubling to speak about, or breathlessly awaited.

Official versus Vernacular Futures

The futures set out in technological visions are different, constructed from imaginative extrapolations of current socio-technical tendencies, buttressed with econometric or cost-benefit analyses, and scaffolded with grand metaphors. Certainly, they can excite enthusiasm, doubt, fear, and even disgust, but they are distinct in quality from lived futures, strictly defined. An analogy can be drawn here between the pair of lived futures and the futures of visioneering, on the one hand and, on the other, the distinction made by Rob Nixon (2011) between "vernacular" and "official" landscapes:

> A vernacular landscape, though neither monolithic nor undisputed, is integral to the socioenvironmental dynamics of community rather than being wholly externalised – treated as out there, as a separate non-renewable resource. By contrast, an official landscape – whether governmental, NGO, corporate or some combination of those – is typically oblivious to such earlier maps; instead it writes the land in a bureaucratic, externalizing and extraction-driven manner that is often pitilessly instrumental (17).

Anticipatory capability and futures literacy are not simply qualities which individuals may have or lack. As they are defined in the first of the forms of lack Sand discusses, they are embedded in authorized practices through which are created portable knowledge objects (like models, cost-benefit analyses, profit forecasts, and so on) that can then become pre-requisites for subsequent actions and strategies (Rip and Kemp 1998). In this way, practices shared by members of bodies which guard futures expertise become ways of building jetties of certainty ahead of the present into the unknown. Such shared practices can therefore be considered technologies, in the sense of artificial means for extending human capability and getting things done. But they are also technologies in the sense identified by Mol (2008), insofar as they are means for realizing particular ends or values that, in the process of their widespread dissemination, may also change what ends or values are prioritized.

In particular, the kinds of practices on which futures literacy relies are ways of rendering a range of different possible outcomes legible by making them commensurable with each other, applying a common measure (such as financial cost) to them. Continuing the mirroring between "official futures" and Nixon's "official landscapes," such practices of commensuration bring wild future possibilities within the purview of bureaucracies in the same way that mapping and statistical techniques made physical territories and demography into governable objects for 18th- and 19th-century European states (Scott 1998). The application of probability-based knowledge practices and the commensuration of outcomes enables future worlds to be constructed as objects of knowledge, sets of expectation values that are framed as risks or benefits. Extrapolating present trends into the future together with the risks and benefits associated with them represents a formalization (Poli 2015, 110) of the future, treating it as the unfolding of a closed system into which contingent events that are unforeseen may erupt, but where such contingent occurrences are at the same time treated as generally reducible to more or less probable ones through the continuing expansion of risk-based knowledge (Groves 2009), "coercively constraining uncertainties into a mold of manageable risks" (Groves 2014, 216).

The analogy between the official landscapes that accompanied the historical emergence of the “social state”, and the "mechanizing" tendency ofofficial futures can be extended further in terms of the geometrical order each extends into their construction of the world and the horizon of their respective visions (Christopher Groves 2017). The meaning-making practices associated with spatial mapping and also with mechanistic future visions both reconstruct the geometrical and topographical order of the world.The geometry of projected, forecast natural futures views the horizon as a receding line, towards which one’s line of sight expands from the point of the present. Robert Romanyshyn (1989) contrasts the linear three-point perspective method developed by Renaissance painters to the methods of medieval painters in which different views of a subject (say, a city) were collaged together without being placed in relation to each other according to general mathematical rules according to which a visual field could be constructed. The dominance of this machinery of representation, he argues, goes on to shapebroader patterns of European cultural change over succeeding centuries. Just as the analogy between landscapes and future maps, drawing on Nixon, proves useful for making sense of the distinction between mundane/profane and elite futures, Romanyshyn’s conceptualization of linear perspective as a materially based cultural technology for making sense of the world gives us a way to think about the cultural power of elite futures. Mechanized futures are geometrical and topographical at the same time as culturally resonant, and as such they reshape perception and experience, and certify particular forms of literacy, of "reading" the future, as more legitimate than others.

A general point to draw from this discussion is that what counts as futures literacy is a matter of social construction, in a similar way to human knowledge more generally. That is not to say that within these processes of social construction, there is no concern with reliability of knowledge (that is, with questions about what forms of socially constructed knowledge survive contact with the world they strive to grasp). But it does suggest that many complex (and some large-scale, and long-term) processes of practice evolution underlie particular estimations of what counts as futures literacy. Through these processes, practices evolve which, by helping to construct futures, also help produce accretions of power, symbolic and otherwise, in the present. At the same time, they often fail tests of reliability, as critiques of the reduction of complex uncertainty to simple risk point out (Wynne 2010). As a result, critical scrutiny is required both of the processes by which a lack of futures literacy is ascribed to some and not others, and of the ways in which dominant futures literacy practices support the social legitimacy and authority of some actors despite being inherently unreliable. That an ascription of a deficit in futures literacy can lead to instances of Sand’s third and fourth senses of a lack of future is an important aspect of contemporary debates over connections between epistemic and environmental and/or social justice – as interactions between, for example, energy industry actors and publics routinely accused of being unable to "think strategically" make clear (Cotton and Devine-Wright 2010).

Profaning Futures Literacy

A variety of strategies can be used to weave together vernacular and mundane, lived futures. On the one hand, there is the question of how elite futures relate to profane or mundane ones. How do members of elite groups involved with anticipatory work reflect on the connections between their professional work and the lived futures (their own and those of others connected to them) linked to it? Often, responses to this question advert to the idea that dominant forms of futures literacy can be alienating and excluding, meaning that encouraging empathic forms of connecting to potential futures is necessary to counteract such effects (Krznaric 2014). This question (essentially one concerned with how to use mundane forms of futures literacy as a means to better communication) has most recently been explored by climate scientists who have articulated their emotional responses to the implications of their own forecasting work for them, their families and the rest of the societies of which they are part, and the biosphere itself (Hansen 2011).

Alternatively, staging confrontations between vernacular and official futures represents a way of subjecting ascriptions of a lack of futures literacy to critical scrutiny (Groves 2015). David Haley, a collaborator with the environmental artists Newton and Helen Harrison, describes their 2007 collaborative map-based anticipatory work *Greenhouse Britain* as a reshaping of the field of play of anticipation, working with visions of landscape change to transform the horizons through which the future is constructed and experienced. In essence, Haley suggests, this work brings mundane, lived futures and elite, forecasted futures together by posing three questions: "How big is here?," "how long is now?," and "who is here now?" (Haley 2011). These questions are intended to unsettle the dominant geometricization of the future through futures expertise which rests on material practices of commensuration. Engaging with people who live in specific places (such as in this case Merseyside and the Lea Valley in London) which will be affected significantly by climate change makes it possible to explore how the horizons projected by futures expertise overshadow other future horizons. More positively, exploring attachments which support lived futures can serve as a means of "profaning" futures literacy that recognizes the contingency and partiality of its dominant forms. In this way, mobilizing mundane, vernacular futures can contribute to what are typically taken to be the goals of futures literacy, namely to "help individuals, organizations, and communities to develop a more sophisticated capacity to choose explicitly among different possible futures" (Poli 2010, 112).

References

Adam, B., and C. Groves. 2007. *Future Matters: Action, Knowledge, Ethics*. Leiden: Brill.

Antonucci, T. C., and H. Akiyama. 1994. "Convoys of Attachment and Social Relations in Children, Adolescents, and Adults." In *Social Networks and Social Support in Childhood and Adolescence*, edited by F. Nestmann, and K. Hurrelmann, 37–52. Berlin: de Gruyter.

Appadurai, A. 2013. "The Capacity to Aspire: Culture and the Terms of Recognition." In *The Future as Cultural Fact*, edited by A. Appadurai, 179–195. New York; London: Verso.

Beck, U. 1992. *Risk Society: Towards a New Modernity*. London: Sage Publications.

Cotton, M., and P. Devine-Wright. 2010. "Making Electricity Networks 'Visible': Industry Actor Representations of 'Publics' and Public Engagement in Infrastructure Planning." *Public Understanding of Science* 21 (1): 17–35. 10.1177/0963662510362658.

Fricker, M. 2007. *Epistemic Injustice: Power and the Ethics of Knowing*. Oxford University Press.

Funtowicz, S. O., and J. R. Ravetz. 1990. *Uncertainty and Quality in Science for Policy*. Dordrecht: Springer.

Groves, C. 2009. "Nanotechnology, Contingency and Finitude." *Nanoethics* 3 (1): 1–16.

Groves, C. 2014. *Care, Uncertainty and Intergenerational Ethics*. London: Palgrave.

Groves, C. 2015. "The Bomb in My Backyard, the Serpent in My House: Environmental Justice, Risk and the Colonisation of Attachment." *Environmental Politics* 24 (6): 853–873. 10.1080/09644016.2015.1067348.

Groves, C. 2017. "Emptying the Future: On the Environmental Politics of Anticipation." *Futures* 92: 29–38. 10.1016/j.futures.2016.06.003.

Haley, D. 2011. "Ecology and the Art of Sustainable Living." *Field Journal* 4 (1): 17–32.

Hansen, J. 2011. *Storms of My Grandchildren: The Truth About the Coming Climate Catastrophe and Our Last Chance to Save Humanity*. New York: Bloomsbury.

Hölscher, L. 1999. *Die Entdeckung der Zukunft*. Frankfurt: Fischer.

Honig, B. 2012. "The Politics of Public Things." *No Foundations* 10: 59–76.

Konrad, K., H. Van Lente, C. Groves, and C. Selin. 2016. "Performing and Governing the Future in Science and Technology." In *The Handbook of Science and Technology Studies*, 4th ed., edited by U. Felt, R. Fouche, C. A. Miller, and L. Smith-Doerr, 465–494. Cambridge, MA; London: MIT Press.

Krznaric, R. 2014. *Empathy: A Handbook for Revolution*. New York: Random House.

Luhmann, N. 1976. "The Future Cannot Begin: Temporal Structures in Modern Society." *Journal of Social Research* 43 (1): 130–152.

McCray, W. P. 2013. *The Visioneers: How a Group of Elite Scientists Pursued Space Colonies, Nanotechnologies, and a Limitless Future*. Princeton University Press.

Mol, A. 2008. *The Logic of Care: Health and the Problem of Patient Choice*. London: Routledge.

Nixon, R. 2011. *Slow Violence*. Cambridge, MA: Harvard University Press.

Poli, R. 2010. "An Introduction to the Ontology of Anticipation." *Futures* 42 (7): 769–776. 10.1016/j.futures.2010.04.028.

Poli, R. 2015. “The Implicit Future Orientation of the Capability Approach.” *Futures* 71: 105–113. 10.1016/j.futures.2015.03.002.

Rip, A., and R. Kemp. 1998. “Technological Change.” In *Human Choice and Climate Change: Resources and Technology*, edited by S. Rayner, and E. Malone, 327–392. Columbus, OH: Battelle.

Romanyshyn, R. D. 1989. *Technology as Symptom and Dream*. London: Routledge.

Sand, M. 2019. “On ‘Not Having a Future.’” *Futures* 107: 98–106. 10.1016/j.futures.2019.01.002.

Scott, J. C. 1998. *Seeing Like a State: How Certain Schemes to Improve the Human Condition Have Failed*. New Haven: Yale University Press.

Sturgis, P., and N. Allum. 2004. “Science in Society: Re-Evaluating the Deficit Model of Public Attitudes.” *Public Understanding of Science* 13 (1): 55–74. 10.1177/0963662504042690.

Wynne, B. 1996. “May the Sheep Safely Graze? A Reflexive View of the Expert-Lay Knowledge Divide.” In *Risk, Environment and Modernity: Towards a New Ecology*, edited by S. Lash, B. Szerszynski, and B. Wynne, 44–83. London: Sage.

Wynne, B. 2010. “Strange Weather, Again.” *Theory, Culture & Society* 27 (2–3): 289–305. 10.1177/0263276410361499.

8 Proactionary or Precautionary – A Debate[1]

Steve Fuller, Alfred Nordmann, Christopher Coenen, Sascha Dickel, Martin Sand, and René von Schomberg

ALFRED NORDMANN: I want to begin with a brief introduction to our discussion. It is an introduction to a debate, which some of us lead in a very passionate way, and some of our readers might find a little arcane and abstract. They might say, "What are these people arguing about? They are all pretty reasonable people, and they all say pretty reasonable things, but they still seem to think that there is a big disagreement on the question of how to best take account and think of the future in our deliberations about emerging technologies, about the relationship of technology, society, and science."

I want to try to highlight where our discussion is coming from, and I have my own position in this. Let me begin with Georg Christoph Lichtenberg, the physicist, author, and aphorist of the 18th century. I draw on him to set the tone for our discussion because he is an Enlightenment thinker. He speaks to us from the century in which our modern political constitutions have been framed.

He made a very famous statement; many of you have read it or heard of it. I think it takes us to the heart of our issues: "I don't know whether things will be better once they are different." When things are different – new science, new technology, new politics – or in other words, when the future comes, who knows whether it will be better or not. I don't know whether things will be better when they are different: "But I do know, they have to be different if they are to be good" (cp. Lichtenberg 1990, 182, remark K 293).

He's being in some sense very careful, intellectually careful, about the future by saying that he doesn't want to presume that he knows things, in effect: I'm not a prophet. I can't claim that we can control the future; I can't claim that I can make the best future or shape it. The future will be what it is. It's a historical process full of contingency. I'm not going to make any predictions and anticipations of the future, but I'm still committed to the future. If things aren't good now and if you want to live in a better world, it can only be in the future, right? We have to work towards a better world, and so we have to make things different even if we don't have predictive or technological control over what is going to happen.

DOI: 10.4324/9781003322290-10

I think that here we already have two different attitudes because there are many people today who do want to know just those things of which Lichtenberg says that he can't know them. They want to know ardently, passionately whether the world will be better or worse once we will have some promised or imagined new technology. In fact, it is safe to say that the cautious attitude of Lichtenberg is not our predominant attitude at the current time. If you open the newspaper, if you go to a science museum, and if you watch talk shows on television, you'll find many discussions about the future of work, the future of mobility, and the future of the human being. People wonder if maybe this or that technology comes or this catastrophe happens: What will life be like and how will we live and will it be good or bad and should we be in favor of it or should we be against it?

This, I think, is a way of thinking that we are now very familiar with: Trying to figure out what the future will be like, especially since we have all those programs and visionary promises of emerging technologies. Those are promises of the ways in which nanotechnology or biotechnologies or artificial intelligence will change our lives. At any given time, that's the big thing right now in that it will change everything, the whole social fabric. So we have a lot of technological promise around us, and thus there's a kind of natural curiosity to say, "Okay, so what is it that is coming to pass here? And how do we relate to it? Do we like it? Do we dislike it?"

This creates also an attitude of a possibility of choosing, preventing, and preparing in proper ways, so we think of the future as a half-baked reality: It's not quite there yet, but it's sort of on the way. All these technologies are on the way; they're coming, and so our role is to be their recipients, and we hope, of course, that we can modulate, regulate, and maybe even prevent some things or encourage other things to happen.

Thus the whole mode of thinking is that maybe we can influence, shape, change, or even design the future. This idea of taking the future as an object of design is motivated by a historical experience. You might say, "Well, look around the world that we live in. It's our world, we made it. It's the product of our actions." So if you look at climate change, environmental pollution, and the like, we say quite rightly that we have made that world. And now the temptation is to say, "Well, since we made it in a careless way, maybe we can make it in a more deliberate, in a more specific, and in a more circumspect way and design it properly." This is not just making the world as a product of our accumulated actions, as a side effect or resultant of what we do, but literally making the future by way of a collective design project.

Now, of course, the more I emphasize or the more I stereotype and caricature this attitude, the less plausible it may appear. And of course, very few people will be so crude and so simplistic and say, "Let's design the future world according to a blueprint." No one quite believes that. But still, there's this general attitude. I want to mobilize Lichtenberg against this, and I want to be the one who says, let's take the future not as an object, not as something that we anticipate, predict, prepare for, and certainly not as something that we choose. There is no choice of doing it this way or that way, and if we do it this way, then this or that will come out.

If we take the future not as an object, let's take it more like a historical category. The future is our task, destiny, or obligation. It's something that we are committed to. We want to have a better future of course. We might always ask ourselves, "how will the people in the future look at us?" And so we are answerable to the future, but we don't actually claim to make it, to design it, or to control it, intellectually or technically. Instead, in the present, we are looking at a kind of extended world that we live in. It is our world that we inhabit. We ask ourselves, is this world sustainable or not? What are the problems in or of this world? Is it a good world? These are the questions that Christoph Lichtenberg might have asked. And we ask, what is maybe not so good about our world? What might be ways to change it? Do we have the means? And thus we try to engage and come up with ways to overcome the obstacles, problems, and challenges in our present.

We hope, of course, that our present actions will actually make for a better world and a good future, but we also know that all kinds of things can happen. They come from nowhere, thwart our plans, and deflect the intended trajectory – a war or an economic crisis or a technological invention that takes everyone by surprise. Suddenly a virus appears, and everything, all our grand plans, are humbled by something that no one had anticipated. Of course, we could have known, but we didn't seriously anticipate, predict, or properly prepare for it, and it's just there, a contingency or externality. That's how history is. We have to accept that as well and deal with it as we incorporate it into a notion of action in the present.

This is my position in the discussion to come, and you might call it a precautionary attitude because I say we are committed mostly to the world that we know, live in, and thus care for. Precaution is an attitude of care. We care for this world, so we want to preserve and sustain it, to make it last. We want to be sure that our children also have a world to live in, and so we have to tend to ours and the problems it faces. We do this with a view towards the notion of a better world, but we don't think we can produce this better world in a systematic way or in the fashion of design.

Part of this idea is also that this is the best way of doing politics – and we will discuss this point today. I would say that dreams are a source of political action in that they offer a diagnosis of the current world and what's wrong with it and how it needs to be improved. The future as such or "technologies of the future" are not a political resource – we might like or dislike the science fiction scenarios in which they appear, but they do not provide us what needs to be done or how to live. I'll stop here for now. I've said a lot – perhaps too much already.

STEVE FULLER: I think this is a very important debate, and in fact, I think I agree with Alfred at least on one point: Politics ought to be organized around this kind of issue that he introduced. However, I would configure it a little bit differently, drawing out the differences between Alfred and myself. Just to start off clearly about this: Alfred takes himself up to be a representative of the precautionary principle, and I, in contrast, am a representative of the proactionary principle, and these are opposing principles. I will outline a bit what I mean by this and what I take to be the difference and then work that into the introduction that Alfred was giving.

While they're talked about as principles of governing risk, the way I see the difference between the precautionary and the proactionary principles is basically their attitudes towards uncertainty, which makes them more relatable to the concept of the future. Coming from the proactionary side, I might be giving a spin to this distinction that people who support the precautionary principle won't necessarily like: The difference of attitude that we're talking about here concerns your default position on uncertainty: Is it a threat or an opportunity?

The precautionary principle, I think, starts on the assumption that it's a threat. That explains the sort of ethics surrounding harm avoidance, for example, and a Hippocratic Oath attitude above all: Do no harm, whether we're talking about people or the environment and so forth. That's because uncertainty – we could talk about the future as being a place where the uncertain happens – is *prima facie* seen as being threatening. Whereas the proactionary principle approaches uncertainty much more, you might say, in the spirit of an entrepreneur who sees uncertainty as a kind of open space to play, as an opportunity space, where things that have not yet been defined can be defined.

With that in mind, I want to move to Alfred's introduction, quoting Lichtenberg. Part of the way in which he framed this, if you recall the way he spoke, is his frequent use of the word "we:" *We*, the collective *we*. We're talking about the kind of future *we* want and to which *we* will be held accountable by the future.

This way of conceiving, perceiving, and seeing things may not actually be the right kind of starting point. In other words, the *we* itself is up for grabs. When we talk about how the world is going to be in the future when there are new technologies and new sciences and unforeseen viruses, the world that we're talking about may or may not be a world in which we – that is to say, those of us who are talking to each other now – will be present.

In other words, we might actually be talking about a world inhabited by different people. Therefore, this idea of *we* needs to be unpacked a little, because if we don't unpack the notion of *we*, if we assume that there's some continuity between the present *we* and the future *we* in this narrative about the future, then we make things a little too easy for ourselves because we're imagining that the future is going to hold us accountable to the moral standards that *we* have now. I think a lot of the moralism surrounding the precautionary principle is basically grounded in that kind of intuition: Future generations will be angry at us for having ruined the environment. This is very much part of the threat, you might say, that is on the horizon and that the precautionary principle is always forcing us to contend with.

But of course, the people living in the future, whoever they happen to be, whether they're literal descendants of us or not, they might have a totally different attitude toward the world they're living in because they weren't part of our world. They live in the future. They're two, or three generations later. They don't know what it was like to live now. And thus the moral standards and normative principles they're going to be operating under are the ones that they can deal with, that they can live with, that they can find acceptable, and within which they can get meaning in their lives.

It is an open question whether that's going to correspond to anything we believed 50–60 years ago. This is how the future becomes a much more open space – the normative space, in other words, is much more open than people who would defend the precautionary principle normally imagine. Moreover, I would say that movements like transhumanism, which are very intimately connected with the proactionary principle, in a way further complicate this picture. If you are a transhumanist, you're imagining a future human being, for example, that might be able to reengineer its body, so that one can live for hundreds of years, which means that the conventional notions of mortality and boundaries around which our entire social existence has been historically organized are blown out of the water.

Other people are imagining that human beings will be able to upload their consciousness into machines and continue in some kind of Silicon form, in which case the environmental demands of such beings will be radically different from the environmental demands of the beings currently inhabiting *homo sapiens* bodies. I think that is pretty obvious: Whether or not any of these future imaginaries ever get to be realized, is of course an open question, and I'm not here to prognosticate or predict anything.

Nevertheless, if any of those things were to come to pass in any kind of way, then there really is a different normative framework in the future, where referring back to what people, let's say in the year 2020, were thinking about as being a livable existence with a lot of meaning, will seem kind of old fashioned and quaint – perhaps the same sort of way in which the so-called modern people look at people who lived in earlier periods, in so-called pre-modern societies. They can understand why people in those places in the past valued the things they did. But, in fact, it remains their world, and we're in our world.

This is where, I think, the issue gets interesting regarding the standpoint of talking about the future in the way that Alfred was talking about, like design and planning. He is absolutely right: In a sense, if you want to talk about what the Enlightenment narrative discourse has been about, it has been a kind of proactionary discourse according to which there is no doubt that science and technology are getting bigger and better, and this is driving the human condition forward in progress.

That has been the normal modernist discourse. However, the kind of world we live in now contends with a transhumanist imaginary, which isn't coming out of the blue. It is reflective of the kinds of technologies that are on the horizon, the so-called "convergent emerging technologies:" Nanoscience, bioscience, and so forth. These things are really putting us on a threshold of a future and future beings who will be inhabiting that space may have their own norms, maybe radically different from the norms of the beings who are living now.

If the future is as open as all that, it becomes really interesting and difficult to determine what should be our normative attitude *now*. In other words, should we be in the business of trying to micromanage the morality or the normativity of the future by saying, "Oh my God, the people in the future are going to regret the fact that we did what we did now?" Can we really talk sensibly in those terms if we're imagining that people might well radically change in the future?

This, then, creates a very interesting and very open normative space, which we have to define. Whatever else is true about this, we cannot fall back on our default moral intuitions that we've been operating with through the modern era. It's really a new space, and I find that very exciting. I do think it really raises very fundamental questions about what life is and what death is and, in fact, what it means to be mindful of future generations. What exactly are we talking about in this kind of open space? Are we doing anything more than reproducing our prejudices and our fantasies from today? This is the kind of challenge that the proactionary principle puts on the table before the precautionary principle.

ALFRED NORDMANN: Thank you very much for your wonderfully vigorous comments which provide a very good opportunity to explain a little bit more the Enlightenment attitude, which I tried to propagate, propagandize with Lichtenberg and maybe even with our great Immanuel Kant in the background. You're quite right, first of all, to say that the way I used the word "we" was too quick. One doesn't have to think very long to know that "we" don't live in a world where there's an agreement about what is the better world.

It is not the case that there is a set of values out there that's universally agreed upon and by which *we* measure ourselves. This would be a kind of essentialism or universalism that none of us is endorsing. It's even a condition of politics that one knows that the *we* is up for grabs: These are contested notions. Part of being in the present means working through this contest of opinions about how to envision or to ask whether this is a beneficial technology or not. Does this actually make our world a better world? Does this actually help people live life more fully?

Even with very mundane technologies – our apps and computers – we can already have disputes whether these technologies actually advance a better world. Thus, this "we" cannot just be postulated as a universal given. I fully agree with that, and I also agree with the second point not to make assumptions about future people, who live in a technologically and environmentally very different world. Future generations will have undergone a historical process that we didn't experience yet, and they will thus have perhaps wholly different sets of values, wholly different sets of debates, and wholly different sets of questions. We shall not presume that there is this seamless continuity, that our actions will be judged from a vantage point in the future with the norms and values that we have now.

This point actually takes us back to some of the early parts of the question about technology assessment and how to take the future into account. I remember a long time ago, when I wrote my first paper about this, this was actually one of the arguments I made: It would be a kind of paternalism if we start projecting our values on the values of others (Nordmann 2007).

This was an objection also against the debates about transhumanism that you refer to: There was a lot of discussion going on a couple of years ago, and it's still going on now, about whether or not we should improve the human mind and the human body and have computer chips implanted in our brains to make ourselves more intelligent and the like: "Should this be done or not, is this right or wrong?" My argument was then and now, "Let's not focus on the future in this way – this technology doesn't even exist as yet, and if it ever were to exist, we should assume that the people who deal with it are very different from us and will have very different values and concerns."

Thus, it is sort of pointless to try to leap over this historical abyss or this gap and to make judgments about this kind of technology and how it will play out. Therefore, I am quite in agreement with you on these two points, but I draw different conclusions.

STEVE FULLER: Can I probe you on those two points? In a way, it is interesting that you brought up the transhumanism debates and what you say about them is exactly right. If you imagine these future transhumanists running the world in some sense, different from us, with these different concerns by virtue of the historical process they will have undergone – let's say fast forward 50 years – then the question will be: Would there be space for people who want to stay in, so to speak, "humanity 1.0?" By that I mean the people who actually do want to carry on the traditional kind of humanistic ideal, that we live a finite number of years in *homo sapiens'* body and that's the end of the story, and all meaning in life derives from that ultimate sense of mortality – will such people be allowed to exist? Could one carry over such a traditional humanist orientation into the transhumanist world of "humanity 2.0?"

The interesting thing is that transhumanists never think about this because they imagine everyone would want to go transhumanist, to take the opportunity to inhabit the world of "humanity 2.0." This is an interesting point, which I do think goes to the heart of the issue about how we are imagining the future.

ALFRED NORDMANN: To some extent, at least for me, it's a question about how we imagine ourselves as being historical subjects. What does it mean to be part of history? In what position does this place us? I think both of us have the sense that there is history and that it involves changing of values, always more than a change purely of technological possibilities, and that we accumulate a past, which also gives us a new starting point for different kinds of futures.

So there is history, but what does it mean to be part of history? I think what you are saying is to recognize that our contemporary *we* or even the plurality of *wes* in our debates is limited. And you say that we can't fall back on that. When we try to look at the larger picture, as you say, the recognition that our own point of view is limited, is actually an opportunity to imagine also other kinds of *wes*, other formations of politics and political debate, and maybe also values and possibilities.

I would say – and this is where I am more conservative and precautionary – perhaps it's true that we can't fall back on our own values to judge proposals for a better world. But, we don't have any other; we only have the ones that we have. This is a limitation, and our challenge is to work with and from this limitation.

On the one hand, this means to acknowledge what we can do. We can only act according to the principles we have. We have this wonderful phrase in German that a conscientious person acts "*nach bestem Wissen und Gewissen*" – "with the best knowledge and conscience that I have." In this phrase, there is an acknowledgment of the limit – "*nach bestem Wissen und Gewissen*" – which means: If I knew more, I could maybe do other things, maybe there would be other possibilities of being, but I only have my way of being. All I can do, in order to take good care of the world, is to do as well, as best as I can, to pursue in good faith what I consider to be perhaps an advance or human progress.

Of course, then things can happen, and maybe some very radical things happen, and something comes indeed often contingently out of nowhere and turns everything upside down. Then we have those moments of opportunity that you have talked about, where suddenly we can think about reconfiguring things, and we have a deep uncertainty about things, for example, regarding the coronavirus right now.

RENÉ VON SCHOMBERG: Steve is able to say interesting things and even things I can agree with. But, on this issue, on precaution versus proaction, this is completely ill-conceived. There are many things which are fully wrong in my view but I will keep it short with a few basic points.

I think Steve actually chooses for his critique a concept of precaution which is extremely weak. It is perhaps a concept of precaution where you assume that people want to freeze us in the current state of affairs, like a sort of conservation principle. But, this is not what the precautionary principle is about. It is not opposed to taking actions, on the contrary: The whole idea of the precautionary principle – when it was introduced under the sustainable development agenda in 1992 and also in the European Treaty – was not as a conservation principle but as a principle for action. It was supposed to function in such a way that the threshold for action for governments is actually lower than it was before.

Before, and in regard to the hole in the ozone layer and this type of challenge, governments had no option but to say: "Look, the science is inconclusive, science is fuzzy, we have no justification for action and don't need to do anything." This is why the precautionary principle was introduced: To lower the threshold for action. Now, this is where the problem starts because then you ask, of course, which action do we then take, in which direction do we go? Here I have to be a little short, but basically, we have to acknowledge that the precautionary principle has some normative elements. If you don't define them, then you can go in all kinds of directions. In modern democracy and in the European Union, we actually do all the things you mentioned. We don't freeze our values as we have them now, we take a deliberative approach. We don't take our society as a collective of people who is a *we*. In a post-metaphysical world, all these things are already behind us. I have the feeling that the current State has a better articulation of these things than what Steve assumes.

Lichtenberg's principle is actually underlying our institutional thinking. But, back to the problem, what type of action should we then take? Arguably, the precautionary principle currently functions as a negative selection criterion by determining what you don't want to do in a certain situation. You can take asbestos as an example: You ban this material because you know that a couple of hundred thousand people will die from it. In this way, the precautionary principle is only negative and does not provide an incentive to pursue certain things. But it could also offer an incentive to go in certain directions, and that is how it is now becoming institutionalized in the European Union.

But of course, this defense of the precautionary principle is not sufficient to address the issue you actually bring up, namely your recommendation of proaction. What does this mean? Now, there have been other people who argued in the same style as Steve for an innovation principle. They provided exactly the same argument and also chose a very simplistic notion of precaution as their foil, believing that it would stifle innovation. This also leads to a question of how to give direction and adopt a course of action.

Again we have deliberate democracy, but secondly, we have a problem giving direction to innovation because there is a pervasive assumption that innovation is intrinsically good, whatever it consists of, because it is beneficial for our economy. This is where, of course, my story about Responsible Innovation comes in: We try to submit science and technology to a deliberative debate in the same way as we do other constitutional elements. We should look at success and failure of our policies regarding science and technology in the same way as we discuss our trade policies or public health policies. Currently, we don't do that but exempt science and technology, and that is a problem if we want to give it direction.

The essential discussion is, therefore, in which direction we, as a deliberative democracy, want to go. How do we make these decisions, and what are the right mechanisms for it? This is the question, and Responsible Innovation is the answer. It is all about institutional change, about changing the ways even of research funding. It is changing our universities; it is changing our economy.

So this is about the transformation of our current society – as such extremely anticipatory. Precaution and action have an intrinsic connection, and so I really have to dismiss this whole contradiction and all the assumptions you make about these principles because we actually are already at the level on which the State has to a certain extent institutionally provided the things Steve calls for. I find it unfortunate that he puts it in the crazy language of "post-truth" and the like. I am sorry to say this, but this plays in the hands of populists.

STEVE FULLER: It sounds from René's remarks, Alfred, that you and I are equally implicated in all of these conceptual confusions, especially you, because you claim to be defending the precautionary principle, or you're falling into a stereotype view of it according to René. But, I want to defend that stereotype view; at least I want to put it on the table for further discussion.

Alfred drew attention to a feature of the stereotype of the precautionary principle which René did not. Whatever may be the normative principles that inform the precautionary principle, they ultimately are about the recognition of *limits.* You can define "limits" in lots of different ways. Yes, we have deliberative democracy in a way that can aid us in defining exactly what we mean by limits, but I think at the end of the day there is a kind of limit notion. That is why there is a tendency, certainly a tendency on my part – and René criticized me for this – to think about the precautionary principle as having a kind of conservation role because of this notion of limit. There is a limited environment, there are limited funds, and people have limited lives. It seems to me that this is always in the background of whatever normative principles inform some particular version of the precautionary principle. So, I put that on the table.

There is another issue that the people who support the precautionary principle need to be clearer about. It concerns the mechanisms of deliberative democracy that René was alluding to: What is this *we*? Who is this *we*? Even if you have a consensus among the people in the room, to whom does it apply (outside of the room)? And for how long does it apply? Is it a binding contract for future generations? To what extent?

RENÉ VON SCHOMBERG: But Steve, this is precisely the point. A deliberative democracy does not presuppose this *we* … .

STEVE FULLER: But you need to solve it in the deliberative democracy. You can't get away with saying "There is no *we*." Otherwise, why should we take the deliberation seriously? It has to apply to something.

RENÉ VON SCHOMBERG: Well, then it would be a very open *we* because anybody can join in; it's across nations; it's all-inclusive.

STEVE FULLER: I thought we are talking about politics where there is a binding decision on the basis of which some collective action is taken. We're not talking about that?

RENÉ VON SCHOMBERG: We also talk about that of course. There you are right. If it concerns a national state where you have to make a decision to accept a ban on asbestos, for example, then I would agree: There is a certain limit, and we have to face the truth. And, the truth is that it causes cancer, and one should be glad that it is banned. But, this doesn't mean that one can also not think about other things. If you put the precautionary principle only in the context of the environmental crisis, then of course it has to do with the idea that you will regenerate the resources which you need for survival. That would be the minimal limit.

But, this is not all there is to the precaution principle. We're all STS people. We know that the technologies evolve over time, and as it evolves, the material means of regenerating natural resources change as well. Our limit changes every day.

STEVE FULLER: I want to bring that up. Can we focus on that point? The reversibility of limits.

RENÉ VON SCHOMBERG: Limits are not carved into stone; all these things are not fixed. If you have new technologies, your limits are different. This is how, of course, the agricultural industry has argued. They say,

"We need more technology in order to nurture 10 billion people, but once we have it, there won't be a problem." Probably they're right; it's possible. These limits are not carved into stone, and neither are our failures.

STEVE FULLER: Well, you need procedures for the limits to be testable and reversible. This is really important, it seems to me.

RENÉ VON SCHOMBERG: Yes, I agree with that.

ALFRED NORDMANN: I want to agree with Steve on this one point that, indeed, what René and I are talking about has a lot to do with limits. You just heard me a few minutes ago saying, in effect: "Maybe our values, our knowledge, and our means are highly imperfect, but that's all we've got, and we're used to work with that."

I'm often referring to our world. A world is something that is given. It is the world that we know, the world in which we can actually make predictions. In a world things have pretty regular ways of coming about, and our world is where we might think, "Okay, what can we do to fix the problems that arise for or within this world?" There are lots of routines within a world that is extended in space and time and I would say that what happens in my familiar world is not really *the* future.

Steve and I, that's where we agree: "*The* future is when there is some other *we*, when there are new forms of life, a new world on the other side. But when we act and deliberate we have to be on *this* side, and whatever we think and do – no matter how deliberate, careful, and forward-looking – it is always on *this* side because we are limited by who we are in the world that we know." For that reason there is a huge difference between saying, for example: If present trends continue, the ocean levels will rise by two meters in 30 years. To me, that's not really a statement about *the* future. It's a statement about the world we live in, the kind of processes that we observe, and the kinds of tendencies we see. But to say that there may be a whole new way of conceiving human existence, questions of immortality, and so forth, is completely different. That would be talking about a future that we can't sensibly talk about, and on this point we disagree.

STEVE FULLER: That's where we disagree, yes.

ALFRED NORDMANN: So, let's take this to the current situation. Steve has published a fascinating pair of papers that he's just recently written in the context of the coronavirus crisis (see Fuller 2020a, 2020b). There is a question of how we interpret this moment because this is one of those moments that are disruptive. Many things don't go the way they have been. We might see this primarily as a challenge or threat, something that challenges us to sustain and maintain a certain way of living, while others see it as an opportunity to reconfigure things.

Steve writes very nicely that we need to abandon certain ideas about what the world is like, what human nature is, and how things are in general. Once we abandon them and put our own values and traditional ways of thinking in a box, so to speak, only then do opportunities arise. He speaks about this as an important moment for politics, and in his paper, he makes an analogy that is very telling especially for Germans: It refers to Bismarck, the first chancellor in the 19th century.

Steve basically says that Bismarck lived in a time when people were actually trying to reconfigure human life, the time when communism and socialism became a strong political force. Bismarck himself was very conservative but he enacted social reforms which introduced something like herd immunity against the radical left and socialism in Germany. Indeed, many historians agree that Bismarck's reforms "protected" Germany from certain socialist, communist tendencies for many years, perhaps even until today. But still, Steve wants to emphasize that Bismarck grasped this moment of opportunity. And he goes on to say that the coronavirus presents a similar moment of opportunity. Perhaps, living with a pandemic, living in the age of pandemics leads us to reconceive the meaning of solidarity and human life.

Of course, I can also see these moments of change. Looking back, historians would say that things are changing, maybe a new world is beginning. This has been a very disruptive moment in recent history. But, for us, right now, I would still say that we act "*nach bestem Wissen und Gewissen*;" that we are still acting in this limited way. I agree here with René that within these limits there are many options and opportunities. We have many ways in which we could proceed. We can have debates about the right ways of proceeding, the right technologies, the right policies, and so forth. Much of that is coming to the floor right now, so we do have a lot to work with.

But, I would not argue that this is a moment of opportunity for grasping and shaping the future. Maybe it will appear that way in retrospect. But looking forward – this is where I support the Lichtenberg/Kant model – we can only act from our limited vantage point: That's the best we have; that's all we've got, and we have to make the most of it.

STEVE FULLER: Going back to the Bismarck example, we are already seeing that people are thinking about the lessons that the coronavirus has taught. People are thinking about the things they have been doing in the past that have been wrong, the things they have taken to be normal that from today's perspective turn out to have contributed to the current pathology.

As we go forward in the future, we will regard certain things that we took as normal as not having been the right thing to have done. This will then be projected into the future so that we avoid doing such things later. In other words, we are involved in a move not only toward a different kind of society; but the internalization of this experience of the pandemic will also involve reassessing what our past has been. We may decide that a lot of things that we took for granted, as having been normal or even good, were kind of bad: We should not have done what we did, and we won't do it in the future; that is part of the reconfiguration.

The reconfiguration is not just about imagining the future to be different from now but also reimagining the past. A lot of things that in the past were thought to be okay or good get reevaluated, and other things that we may have underestimated will perhaps be evaluated positively. In short, the crisis may result in a complete *Gestalt* switch of our own relationship to history – not just the future. That's how I would frame this.

CHRISTOPHER COENEN: I will try to be a bit provocative: We have an elephant in the room, and it is capitalism; and that means, we don't have a *we*. At the current COVID-19 moment, this is quite clear: Everyone can see whether people who are working in so-called system relevant jobs are fairly treated or paid; and most often, they are not. At the same time and without discussion, some areas of the economy are exempted from lockdown measures. I want to ask, in particular, René: Where is the *we* in Europe when we have autocrats in some European states, including in the EU, and populists everywhere? And what does it mean, for the old *we* of progressive humanity, including early transhumanism, if the people who, besides Steve, still believe in a grand future of humanity are the digital capitalists in California? I think it's not a question about what *we* do or not do, because there is no real *we*.

I certainly think that one can believe in the future and in technology, but it's a question of political systems first. I don't like the Soviet ideology, but at least it was based in the hope that certain benefits of technology are justly divided. This is something that I like in René's point of view: He always brings the political dimension to this discussion. But we have no basis at the moment anymore to lead this discussion. If we do not save our democracies first, we will have no democracies and no civil society in the future that can steer techno-scientific developments.

SASCHA DICKEL: I would like to catch up on the problem of the *we* because, as Christopher pointed out, there is no *we* that is able to decide on the way that we should collectively take to approach the future. But, then you again argued that *we* need to find a solution to all these kinds of problems, so you again assumed *we* as a collective entity, that can in some way approach the future and is confronted with all these problems in need of solutions. I think this is exactly one of the issues that makes me skeptical about the capacities of deliberative democracy to deal with the radically different attitudes in the scientific community, with the radically different ways of approaching the future.

As you mentioned, Steve, the proactionary principle corresponds to an entrepreneurial way of approaching the future. But there are also legal, religious, and scientific ways, so we have all of these different spheres of values in society with radically different ways of dealing with the future. For example, in the economy, the future is a sphere of opportunity that can be grasped, that can be defined, created, and designed. In politics, the future is rather something that needs to be taken care of and that needs to be governed. How can we deal with these different, radically different approaches? Deliberation is at least an attempt to deal with these different ways. But, when we undertake it as a political entity that assembles publics and starts exercising deliberation, then the future is again turned into a subject of politics and care.

So, when you engage in a deliberative exercise and you assemble different segments of society, people from the business sector, from religions, civil society, etc., then you do it under the umbrella of politics. The future becomes a political issue, and maybe you lose the authority of deciding the way of approaching the future because the participants of this discourse need to subscribe to the political imaginary of the future, the imaginary of careful governance.

ALFRED NORDMANN: Let me express my puzzlement towards both Christopher's and Sascha's point because, yes, the question of the *we* is a big political issue and one shouldn't be too quick to assume that there is such a thing, but is this the problem that we are talking about? Is this the problem, whether one can somehow bring this *we* together so that we can arrive at a common conception of the future or of a just world that we collectively pursue? This is not our problem. What is the future, is it an object of such kind of consensus or dissent? It is the result, in the end, of all the things that all kinds of people do, think, and say, and of all kinds of contingencies. No matter how strong a collective will is, we are not making the future according to this particular will. The historical process results from the accumulation of many, many decisions and actions. For me, politics is not about agreeing in a deliberative setting on what the future is, but it's about the whole process, all the different practices that compete, and some of them are more innovative, some of them are more conservative, and so forth. In the end, something comes out, comes to the fore, as we muddle through, and we have to hope, everyone hopes on his or her own terms, that it is a move towards a better world.

When I look at the corona situation right now, I see primarily all the questions that Christopher is raising, and of course, you could say – this is a metaphor that Steve and I might share – that the corona situation is like a stress test to our societies, as we know it from the financial crisis where we had stress tests of the banking system. So now we have a condition which brings out the weakness and strength of our political systems, our healthcare systems, and the like.

So we have a stress test, but it is interesting for me to see that it is working out in an extremely conservative way, that we witness a rise of the authority of science which is almost unheard of. Only half a year ago we talked about science losing credibility: No one wanted to trust scientists anymore. Now we are in the grip, in the hold of science and not just of any old science but the science of population control, that is, 19th century statistical science having the loudest voice.

This is very interesting in regard to the idea that we have to somehow tend to the world by playing it safe, by taking the most conservative kind of science we have. Of course, we all hope that we can get beyond that and to reclaim a more proactionary way as René was talking about it.

RENÉ VON SCHOMBERG: I want to come back to Alfred's quote of Lichtenberg because I think it also assumes an element which is essential for modernity – the notion of progress. Exactly in the way Lichtenberg expresses it, you anticipate the possibility of improvement, but you don't know yet what it will look like, or how it will be shaped. This is precisely the point, and I know this notion of progress is not very popular these days.

Now it is popular to talk about post-truth and associated notions. But let's face it, also in our discussion now there is the assumption which we actually seem to agree on, that you can actually make progress. But what does this progress mean? How do we get to another level with respect to deliberative democracy which, in my view, still has to evolve? This is why I say that Responsible Innovation, for example, is a radical socio-political program for institutional change. It is nothing less than that. It means a radical reformation of the economy, and actually, corona is indeed an opportunity for this.

Corona clearly demonstrates that for its solution it cannot be left to the economic sector. Christopher rightly points out that ideas like progress are not so popular anymore, so people actually like anti-deliberative solutions, and unfortunately we witness that in Europe and elsewhere. This leads to a societal division. A crisis is used by politicians in all kinds of ways: Some exploit it as I tried to do it, others in their way to install a dictatorship. We will see who wins this battle because it will be a battle.

I think you cannot conceive a deliberative democracy without a notion of progress, without the notion of continuous institutional change, and without the notion of a very open society where, indeed, you don't have a *we*. Dictatorships have a *we*, but I think the European Union doesn't have a *we* according to its treaties. So, by defending Europe, we are actually defending something which is subject to continuous democratic improvement, a process which unfortunately some have put in doubt, and some even anticipates its disappearance, which would imply a reactionary regress – so it's really a strange society where we live in nowadays.

SASCHA DICKEL: The point I am going to make is a little bit unrelated to the discussion right now, so maybe we can come back to it later. Both of you made an interesting point, and I totally agree with it: We usually only talk about the future explicitly when it is already problematic in some way, when it is already uncertain. It's a bit different when we think about it in a more mundane way as something that is not talked about, that is not brought up in discourse.

When we think about, for example, tomorrow or the next week, then surely this mundane future is an object of design. We treat it as something that can be planned, that can be arranged, that conforms to our routines, and that doesn't surprise us normally. Hence, I asked myself, how do we know when the future becomes problematic, when our routines won't fit anymore, and when we can't treat the future as an object of design and planning? Is it a matter of time spans or is it simply one, two, or three generations away?

Building on these questions, is there some kind of epistemic technology that can be applied to everyday life or politics, so we know the point at which we are no longer able or shouldn't be tempted to treat the future as an object that can be designed or planned in a routine way?

STEVE FULLER: I'm just going to respond to this last point because it is pertinent also in regard to the coronavirus. In the two papers of mine that Alfred referred to, I talk a lot about how you might put a proactionary spin on the corona pandemic, and how I think that might happen. I do think that if you actually look at the way in which it's being dealt with on a day-to-day basis, one thing that really strikes me, and maybe it strikes you as well, is how much the concerns are focused on the present.

What I mean by that is the whole idea of whether we're prepared mentally – and by "mentally" I mean psychologically and not just conceptually – to imagine a future rather different from the present. What I would point to is the fact that everybody is preoccupied with keeping everyone alive. I know that sounds like an obvious point, but it is very striking.

A different approach was taken in the UK and in Sweden, where people started talking seriously about "herd immunity" – and herd immunity presupposes some people are just going to die, and the only question is what the long term will look like. That's one way of naturalizing the virus: The virus becomes something that is tolerable within the population that stays alive as a whole, but that means that some individual people will die, and these will mostly be the people who are in some sense close to dying anyway. This is the ethics behind herd immunity.[2] Basically, what you're doing is you're making the vulnerable slightly more vulnerable. They disappear, and then you get this new world. Then there will be a new kind of business as usual.

Now when this idea – and this is a familiar idea from public health and epidemiology – was floated in Britain and in Sweden, there was an enormous amount of pushback. Everyone is concerned these days about saving everyone‘s life, so one of the consequences is that we don't have a clue about what no-lockdown looks like. There is no image of the future. There is no image beyond the coronavirus pandemic because everyone is so preoccupied with keeping the people around alive and death rates controllable. The future has been suspended in the way in which the virus is being handled on a day-to-day basis. That is interesting because at a practical level, it shows the unwillingness of people to talk about the future in a relatively detached way. I just put that on the table. To me, it shows a very precautionary attitude, more than I would have imagined.

MARTIN SAND: Coming back to the original contrast between the precautionary and the proactionary approach, what is striking to me is that we have seemed reached a point where we say that precautionary doesn't necessarily mean being entirely inactive, while proactionary doesn't mean anything goes.

In the end, therefore, is this only a matter of emphasis? Is this a matter of how strongly we emphasize either that the future is a threat or that it is an opportunity for the advancement of society and for humanity? We have discussed this on a very abstract level, but would these two views about the governance of technology in our society finally come to the same conclusions about practical governance?

Would the proactionary principle come to the same conclusions that René pointed out about the rejection of asbestos? I'm not sure about that. And I'd like to add another observation about this: These two approaches seem to be motivated, on the one hand, by how we evaluate the future. On the other hand, there is a different emphasis in the evaluation of the present. Particularly when I think about transhumanism, I think about people who find the construction of the human body very insufficient. The fact of our own mortality is here evaluated negatively, so the motivation to propose proaction is not only that the future might entail a lot of opportunity and possibility, but also that the present is so insufficient and unbearable. So we have two diverging sources for the difference in emphasis. One is the perspective on what the future holds, and the other is the evaluation of the current state of affairs.

AUDIENCE MEMBER [MELANIE JOHANNING]: There is a general question concerning some points mentioned by Steve about how to judge the future. As far as I understood, he said that our norms will change. The conclusion would be that I don't know how the norms will be in the future, so future generations might not judge my action right now the same way that I do. This leads to the question how should I behave when I see that we are doing things that, from my point of view, I know not to be good? I know, for example, that some actions in my private life might be bad for the climate. If norms will change, why should I act on my current set of norms? By which norms would you judge which opportunities to pursue? After all, in our daily life and also in technology development, there are so many decisions and many ways one could go. So which way or which norms should we take then to judge and make a decision?

STEVE FULLER: You are raising two sets of questions. Let me begin with the second one. I think that obviously we now make normative decisions about the world in which we live. There is nothing particularly problematic about that. But as for the standing of those decisions – this perhaps goes back to the earlier point about asbestos – they should be made in such a way that they are reversible in principle under certain circumstances. It seems to me that democratic accountability, as we normally understand it, builds that in, at least by way of regular elections. In a sense, the electorate is forced on a periodic basis to actually ratify or reject what the policies have been up to that point. In fact, that is part of the genius of democracy: Binding decisions can be reversed on a periodic basis, and the people are forced to examine them again.

I don't think there is anything particularly wrong with the norms we have now when making decisions about technology, weapons, and the like. The question is, how much binding force are they supposed to have in the future? For the most part, they won't have much binding force, they'll be reversed at some point or revised or looked at again in some sense, so I'm not too concerned about that. The only concern I would have would be if someone holds them to be irreversible. So that answers one of the concerns in the question.

RENÉ VON SCHOMBERG: But Steve, you are now actually articulating an element of the precautionary principle. In fact, you are defending it, because one of the reasons the precautionary principle was introduced, was to prevent that irreversible decision will be taken.

STEVE FULLER: So, you'll be able to reverse the ban on asbestos? That's in principle up for grabs too?

RENÉ VON SCHOMBERG: Yes, it's possible. I don't think it will happen in the next 10 years, but in theory, it can happen. You might think of a new technology which solves these issues. As you say, it's of course always reversible since reversibility is institutionalized.

STEVE FULLER: Right, but the option is not enough, the form of institutionalization matters. It was Karl Popper who made the point that reversibility is not just an option to have, but rather that one should be forced to regularly consider reversing a previous policy decision.

RENÉ VON SCHOMBERG: But these are things we don't need to argue about now; they have been institutionalized for decades. That's a given.

STEVE FULLER: Some things are given and some aren't. The point is that I don't think every policy decision is quite as institutionalized as you make it out to be. Okay, that would be one point. Here perhaps our difference is one of degree rather than of kind. This goes to the question that Martin was raising about just how far apart we are on things. But one thing that defenders of both the proactionary and the precautionary principles need to discuss is the issue of the negative externalities of decision-making, namely the people and things that are affected though they are not actually part of the decision-making process.

I think one of the problems with the proactionary principle, especially in the hands of certain libertarian transhumanists, is that they don't have a conception of negative externality. They don't see the implications of their decisions on other people who aren't involved in the decision-making. This is a weakness in the political vision of some people who adopt the proactionary principle, namely, that negative externalities are not alive enough for them. They seem to think that the people who make decisions are the ones who bear the consequences, end of story. I guess you and I would agree on that, René?

RENÉ VON SCHOMBERG: Yes, and I actually think we are coming closer than ever before. I have the feeling we actually agree on the notions of deliberate democracy. I only think that most of what you want is already institutionalized, but there, of course, we may disagree: It's all about how we give direction to our innovations and how to make progress. How do we do that institutionally in a deliberative democracy? Of course, we are here together and might actually have similar ideas as well, but with respect to progress, this presumes that we have a socio-political program which makes it possible to go there. In the current political situation, this is questionable, but there we go.

STEVE FULLER: Just one more small point going back to Martin – he was putting his finger on whether you think about the present as something that we have got to maintain, or we need to make incremental progress? Or, do you think that the present is radically bad or in some sense much worse than it really could be?

RENÉ VON SCHOMBERG: This is what I am just trying to say: We have to assume in modernity, and this is what I defend, the idea of continuous progress.

STEVE FULLER: But where are we starting from? Pretty good or woefully bad?

RENÉ VON SCHOMBERG: It's neither good nor bad. This is what we all say: The *we* and the people change all the time; values change all the time. Thus, how we understand the idea of making progress also changes all the time, and this is why we have to keep on talking. But, it does assume a certain institutionalization of progress. If 90% of people want to bring back the death penalty in the UK, for example, I would consider this as a regress in the real sense of the word.

ALFRED NORDMANN: Okay, let me just jump in one more time. Sascha was asking, "What is the criterion? How do we distinguish the kind of mundane future that is part of our daily lives, and that encompasses only tomorrow or the day after in the normal course of things, and that other kind of future? Is there a criterion for distinguishing them?" Martin asked a similar question by saying "Are our differences of opinion only a matter of emphasis?" I want to get back to my basic point, and I think there is a criterion. The criterion is precisely whether or not I have to take a leap to imagine a different world.

What is proposed as a historical future of humankind and also what is imagined as a future in science fiction novels invites us, more or less explicitly, to take this kind of leap. As we are invited to take this leap, we are also reminded of our limitations since we cannot really imagine what it is like to live in this future, and we can't really imagine the kind of subject or person who inhabits this other kind of world.

If there is this radically discontinuous different world, it will also be populated by people different from me. This is exactly Steve's point. And so when I'm invited to take that leap, I am overtaxed and have to decline and say something like: "I know that my actual given world is a limited world, a contingent world with accidental features, and it is not the end history. It cannot and perhaps should not be preserved, but it is all I have, and that's what I am working with." This is my criterion, if you want to call it that.

Now the question is, what is happening in this world? Let me come back to the corona situation because it is on everyone's mind and also because Steve said something absolutely brilliant about it. He said that, indeed, what we're observing is a very precautionary attitude, and this actually makes your point. If you look around, you see that everyone is trying to save lives and that people have in some sense suspended the future. There is no imagination of the future, so what is happening is exactly what you've been worrying about. Here I have a different view, but it's also different from René's.

I would say the first thing about the coronavirus is that it's been very humbling. We thought that things were moving forward in all kinds of ways, and we were all worried about things like climate change and sustainable development, and suddenly there is a different kind of vulnerability, an entirely different kind of thing that no one properly anticipated. It derailed everything, and that's, at first, a very humbling experience when it comes to our view of being in control, of knowing how to move forward.

Therefore, I'm not sure whether the idea of progress, as René proposed it, is still very prominent in our situation right now. In fact, I find in my engineering students the predominantly precautionary attitude which Steve has identified. When I remember my own student days, I was part of a group where our parents were hoping that we would have a better life than they did. I think this was true for our parents, our grandparents, and our great-grandparents: They were always hoping for a better life for their children in the future and so – this goes to your question, Steve – they thought of their present as a deficient present. When I ask my engineering students today, even the ones who might have fancy ideas about new technologies, they don't believe that their children will have a much better life than they do. I think in that sense this pathos of progress doesn't exist anymore.

In the end, however, I also wouldn't agree with you, Steve. You said that we are getting so fixated on sustaining life that we don't have a vision of the future. I would say that it's not really about saving lives. It's about keeping the infection rate low, leveling and flattening the curve – and this is a management problem. We're tending to the management of the infection rate. Now, when we look at the ways to actually manage this problem, we are finding out that in fact there are many different approaches, that we have many different kinds of knowledge to draw on. We notice that we can have knowledge of population control which keeps us separate and locked down. Or maybe we can utilize a hodgepodge of local knowledge stitched together for protection against infections and for finding ways to incorporate the existence of the virus into our daily lives.

I think there is a very lively debate right now about the different options we have. It's not like we were left without any options. And this is why I would say that we are tending to the present but not just in the conservative way of maintaining and preserving it. We have alternatives to discuss, there is contestable politics about how to manage the situation. In order to seek out the politics of the situation, we do not have to take the leap into a different way of thinking, a different kind of way of living, or a different set of values.

CHRISTOPHER COENEN: I think the notion of progress you mentioned brings matters to a point. The basic idea for all of us is that we want to see things better in the future, right? This is traditionally the notion of progress, and we might diverge on the point of how things can be better.

But what I would really like to point out is that this is not the situation we are in anymore. We are not in the 1970s or 1980s when both sides, capitalism and communism, were on the side of progress. Now we have a different situation also in Europe: There are 15%, 20%, and in some countries 40% of the population who think, for example, that they are in an information war, often believing in right-wing or other conspiracy ideologies.

What I wanted to point out through my earlier provocative statement is that we are discussing a world that is gone. What I learned from this discussion is that the differences between all of you are not so big. I heard that the imagination of a better world is often missing – but before we can recover this imagination, we have to defend democracy, which cannot be taken for granted and is under attack.

Notes

1 This discussion took place on April 28, 2020, in an online format in front of a student audience at the Technische Universität Darmstadt.

2 To be sure, herd immunity can be achieved by way of vaccination as well. Steve Fuller is here referring to the early debates in spring and summer of 2020 when herd immunity was achievable only by letting the virus run rampant.

References

Fuller, S. 2020a. “A Post-Truth Proactionary Look at the Pandemic,” *Postdigital Science and Education 2*: 551–555. https://link.springer.com/article/10.1007/s42438-020-00124-5

Fuller, S. 2020b. “Fourth Order Thinking About the Pandemic: A Transhumanist Challenge,” *Social Epistemology Review and Reply Collective*. https://social-epistemology.com/in-and-beyond-the-era-of-covid-19/#fourthorder

Lichtenberg, G.C. (1990). *Aphorisms*, London: Penguin Books

Nordmann, A. (2007). ‘If and Then: A Critique of Speculative NanoEthics.’ *NanoEthics*, 1, 31–46.

Part III

Exemplary Explorations

9 Prototyping Futures – Towards a Hermeneutics of Artefacts and Technologies

Sascha Dickel

The chapter aims to problematize and expand the hermeneutic approach to the future. It emphasizes the relevance of material artefacts and technologies in communicating and constructing technological futures and introduces a hermeneutic approach to the technological design of present futures by analyzing prototypes. I suggest that prototypes themselves are media of communication in which anticipations are inscribed and made readable (Rip 2009). An analysis of prototypes allows reflexive researchers of futures to expand the focus of hermeneutic interpretations.

After a short discussion of the hermeneutical approach to the future and some methodological considerations of studying artefacts and technologies, I will formulate suggestions on the epistemic and social function of prototypes. The chapter suggests that prototyping turns the present into an object of design instead of giving access to *the* future.

Hermeneutics of the Future

How may we get access to the future? Following the modern ontology of time, the answer to this question is "never".[1] The future is not an object that modern subjects may investigate like any other aspect of the world, because (at least for them) the future has not yet arrived. Luhmann famously argued that the modern "future cannot begin" (Luhmann 1976), because we will always be anchored in a specific present. The *future presents* ahead cannot be known. The more modern subjects accept and reflect this aspect of their temporal existence, the greater the urge to construct *present futures*, constructions of how their future may be, will probably be, or should be (Adam and Groves 2007; Koselleck 1979). "Wherever we care to look we cannot fail but notice that the contemporary capacity and competence to produce futures is phenomenal" (Adam and Groves 2007, xiv). Present futures *represent* the absent future presents.

The proposed hermeneutic turn in technology assessment (TA) and responsible research and innovation (RRI) (Grunwald 2014; see also Nordmann 2014) uses this temporal paradox as a starting point for reflexive studies of technological futures. This hermeneutic approach does

DOI: 10.4324/9781003322290-12

neither aim to predict the future nor does it aspire to construct possible, probable or desirable scenarios. Rather, it acknowledges the "immanence of the present" (Grunwald 2006) inherent in any construction of the future – be it a fictional story, a scenario, or a prediction. The goal of hermeneutics of the futures is to understand the meaning of such constructions for the present.

This hermeneutic approach links to a longstanding discussion about expectations, futures, and temporality in the social sciences and humanities (Bell 1971; Brown and Michael 2003; Gumbrecht 2001; Hölscher 1999; Kaiser 2015; Koselleck 1979; Luhmann 1976; Selin 2008). It shares with these academic works the epistemological and methodological assumption that *the future* (in the sense of the future presents) cannot be known. It therefore urges TA and RRI researchers to focus on *discourses on the future,* to analyze on the one hand the content of techno-futuristic narratives, and on the other hand the social forces these generate futuristic narratives. "The communication of techno-futures constitutes an intervention in ongoing communications. It can lead a debate to take a turn and influence decisions, possibly even independently of how consistently, plausibly, or scientifically well founded the respective futures are" (Grunwald 2014, 287).

In Grunwald's (2014) programmatic paper on the hermeneutic turn in TA, he distinguishes the analysis of the *contents* of present futures from an analysis of the *construction* practices of futures. A similar division of (research) labor is articulated in a collaborative position paper on the hermeneutic approach by the German TA community (Lösch et al. 2019). Its authors distinguish two analytical dimensions: The analysis of futures as objects of study ("society in futures") on the one hand, and the processes and constellations that govern the production and circulation of futures ("futures in society") on the other.

Grunwald suggests that we encounter present futures as "stories and images" (Grunwald 2014, 279) of the future. Thus, he conceives futures as products of language and suggests that "techniques of social scientific discourse analysis are methods suitable" (Grunwald 2014, 287) for the reconstruction of futures. "The role of language is fundamental in discourses on the future, because *the future does only exist in language"* (Grunwald 2006, 56).[2] I agree with Grunwald that present futures exist as products of communication. Language, however, is only one possible medium of communicating futures. It is for sure a very important medium because without language, complex constructions of the future would be impossible. Hence, when we talk about expectations of the future, we are indeed often talking about narratives of the future that are expressed in language. Because of the importance of language for all kinds of social constructions, the value of an analysis of futures in the medium of language can hardly be contested.[3] However, current academic debates on materiality suggest that the focus on language needs to be complemented by approaches that emphasize also non-verbal media.

Materiality of Futures

I take Marshall McLuhan as a starting point for my reflections on materialized futures. In the 1960s, McLuhan urged scholars of media to shift their perspective. According to McLuhan, the culture of societies, the various ways which enable them to understand their world and themselves, is very much shaped by the media technologies it deploys (McLuhan [1964] 2003). Instead of studying the content of media studies should focus on their materiality. McLuhan's approach has often been accused of technological determinism. However, this reading of McLuhan is rather one-sided and ignores that McLuhan incorporated not only material technologies but also the social practices associated with these technologies into his analysis of media (Balka 2000). "As an interactive node, a medium as a mode of producing, transmitting and/or distributing communication has its shaping power on the everyday world, but it is also shaped by the priorities of that very same everyday world [...]. [It] does not determine certain events, it interacts within them, opening up possibilities, amplifying programs, undermining resistance to change and itself is changed and directed in the process" (D. Theall and J. Theall 1989, 49). Seen this way, McLuhan's work anticipated later approaches in science and technology studies to reconsider technology's active role for social orders (Balka 2000).

Before McLuhan, the study of media was dominated by approaches focusing on content. McLuhan, however, emphasized the importance of the materiality of media. He claimed that not only *what* is communicated is an important object of study, but also (and even more) *how* – by which material means and practices – something is expressed. McLuhan suggested that scholarly fascination with the content of media obscures the understanding of media technologies that enable, structure, and shape the production and reception of content.

By analyzing news on television, for example, we might focus on the presentation of a particular political conflict. We might analyze how a specific news program presents a conflict, how conflict parties are portrayed, what gender roles are implied in the narrative, and what values are expressed in the comments of journalists. This would be a *hermeneutics of content* that is interested in the ways in which actors interpret their world. This approach is based upon a phenomenology interested in human meaning-making, but not so much "in the *technological tools* of meaning-making practices" (Tripathi 2016, 147). By contrast, McLuhan's focus of analysis would exactly be these tools of meaning-making. He wouldn't, for example, investigate the content of a particular news program, but the ways in which television as a media technology transforms and redefines the ways in which news are produced and consumed (McLuhan [1964] 2003). The famous quote of McLuhan, "the medium is the message" refers to these reconfigurations of meaning with the introduction of new media.

How deeply reconfigurations of media might change social practices can be exemplified by Niklas Luhmann's analysis of the transformation of communication by technologies of writing. Compared to face-to-face interactions, written texts opens new possibilities of communication, but also generate new restrictions. Writing makes it possible for communicative acts to transgress time and space. Memory can be externalized, information can be expressed in more complex ways. Writing emphasizes *what* is said, while the social relationship between speaker and listener is suspended and slides in the background. At the same time, writing makes it impossible for the reader to react to the writer directly and instantaneously. The mutual taking of turns in face-to-face interactions, which allows for an exchange of roles between speaker and listener, is impossible in the relationship between writer and listener. Decoupled from the social boundaries and sensibilities of interaction, writing enables novel and more complex forms of critique. Writing also leads to new risks of communication. It becomes uncertain, if the message reaches the intended reader, and it becomes nearly impossible to limit the scope of possible readers. Different from speakers, writers must assume that an unknown audience might further distribute the message and react to it in unexpected ways (Luhmann 1997, 249–90).

The focus on materiality, so prevalent in the works of media theorists like McLuhan or Kittler (2006), has become prominent again in recent academic discourses (Barad 2003; Braidotti 2013). Years after the "practice turn", a "material turn" in social sciences and humanities promises to overcome the analytical limitations of the "linguistic turn" that still informs various paradigms, theories, and methods in the social sciences and humanities (Lemke 2015). Neo-materialists draw attention to the fact that materiality matters. Hence, if we want to understand futures, we might be well advised to expand our hermeneutical analysis towards an analysis of the meaning of material objects as well as their roles as "*technological tools* of meaning-making practices" (Tripathi 2016, 147).

When researchers focus on the making of futures, they often investigate the *actors* that construct futures and the social *contexts* that enable and shape the production and reception of futures (Grunwald 2014, 286–87). Futures, however, are not only shaped by the intentions of individual and collective actors or specific societal forces (e.g. political, economic, legal, scientific, or religious logics and constraints). There are also certain routines, scripts, and methods at work, that influence how futures are constructed. One might call these the *technologies* of "futuring" (Cornish 2005). For example, if futures take the form of texts or are turned into movies, the technologies of writing texts or producing movies will shape the outcome. They determine what types of forms and contents are possible in the first place and may influence which kinds of actors are involved in which roles. By studying the entanglements of practices and materialities assembled in these technologies of futuring, one gains further insight into the "apparatuses" (Agamben 2009) of futuring. Just like the content of specific technological discourses tells us something about our present, these practices and materialities help us understand our contemporary culture of futurity and innovation. The making of artefacts (as media of present futures) requires other technologies of production compared to texts or films. It assembles other actors and is shaped by other scripts and routines.

The empirical case of this chapter is the construction of specific kinds of artefacts: Prototypes. What does it mean to articulate futures via the *medium of prototypes*? And how does prototyping work as a specific *technology of futuring* – as a practice that explores something that is not yet there, that has not yet materialized in a form that is considered finished and complete?

Prototyping the Future

The contemporary usage of term prototype refers less to the past (to an original form), but more to an "objectivation" (Berger and Luckmann 1967, 49–50) of present futures. Prototypes are objects "in which an idea is manifested in different stages of development—in part only in its selected properties and components" (Gengnagel et al. 2016, 5). Prototypes are "socio-material devices for ordering the future in the present" (Wilke 2010).

Practices of prototyping are nothing new. Prototyping *avant la lettre* "has always existed and probably, for most of human history, has been more important than it's [sic!] opposite, orderly science and planning" (Guggenheim 2010). In his seminal work on *thing knowledge*, Davis Baird argues that "[t]hought and design are not restricted to processes conducted in language. And working with models and prototypes is not more primitive [...] than working with words and equations" (Baird 2004, 149). In contrast to the long neglect of prototypes (both in innovation research as well as in humanities and social sciences) there is an increasing academic attention to this inherently unfinished object (Guggenheim 2010). The social studies of prototypes so far focused primarily on their function in practices and processes of innovation (Suchman et al. 2002; Vetter 2011). Some have begun to analyze the function of prototypes to materialize what is not there yet (Buchli 2010), their role in specifying, evaluating, and demonstrating a technological future (Schulz-Schaeffer and Meister 2017). If "the medium is the message," what could be the message of prototypes? How do prototypes establish connections between the present and the future? What kind knowledge do prototypes convey? What social function does the presentation of prototypes accomplish?

The emphasis of the following inquiry is the reconstruction of "technologically mediated social orders" (Okabe and Ito 2006, 3), thereby drawing attention to the "technological condition" (Hörl 2011) of social life in our contemporary world. While society has always been co-constituted by technologies, the notion of technological mediation emphasizes the growing importance of technologies in all social spheres, especially the new social configurations and relations that emerge in the context of digital media technologies. My theoretical investigation is informed by several empirical case studies on prototyping in maker-spaces and hackathons by means of document analysis, participant observation, and interviews. Nevertheless, this chapter is not so much interested in *specific* prototypes, settings of prototyping, or the situated images of the future entangled with them.[4] It is rather interested in the question what it means to access and communicate futures by means of prototypes as *media of communication* and prototyping *as a technology of futuring*. Hence, the empirical analysis remains in the background of the chapter. Its results are rather used to inform the theoretical argument.

Prototypes are objects of desire. They symbolize a future, which has not yet arrived, but can nevertheless be specified, evaluated, and demonstrated (Schulz-Schaeffer and Meister 2017). Prototypes are not stable instruments that can be used in routine practices, nor are they products ready for the market. In laboratories, design studies, and makerspaces, prototypes are embedded into epistemic practices, always in the process of being reshaped and turned into another materialization of the future. Prototypes are therefore a prime example for "epistemic objects" (Knorr-Cetina 2001).

Knorr-Cetina's differentiates epistemic objects from *equipment* (as defined by Heidegger) and *commodities* (as defined by Marx). Objects function as equipment ("Zeug") if they are means to an end. In its instrumental usage, the object itself becomes transparent (see also Star 1999, 381). "Equipment becomes problematic only when it is unavailable, when it malfunctions, or when it temporarily breaks down. [...]. Thus Heidegger characterizes objects of knowledge in contrast to instruments in terms of a 'theoretical attitude' that entails a 'withholding' of practical reason" (Knorr-Cetina 1997, 10). As instrumental equipment, objects are interesting only in terms of what you can do with them and what they are used for. Something similar applies to commodities. Following Marx and Baudrillard, Knorr-Cetina argues that as a commodity, an object is "not valued for its intrinsic properties, but rather for what it buys – status, relationships, other objects, and so on" (Knorr-Cetina 1997, 11). As commodities, objects are reduced to their (material or symbolic) exchange value. Instrumentalization and commodification are characteristics of the object relations of industrial modernity. In contemporary knowledge society, however, a new class of objects is gaining in relevance: *Epistemic objects*.

The concept of epistemic object builds upon on Rheinberger's (1992) notion of "epistemic things" used in scientific research, which are open and question-generating. In contrast to technological instrument that are ready-to-hand, stable and unproblematic, epistemic objects are instable objects of problematization:

> They are processes and projections rather than definitive things. Observation and inquiry reveals them by increasing rather than reducing their complexity. [...] [W]hat an object is at present to some degree depends on how its future develops. Building upon Rheinberger's ideas, I want to characterize objects of knowledge ('epistemic objects') in terms of a lack in completeness of being that takes away much of the wholeness, solidity, and the thing-like character they have in our everyday conception.
>
> (Knorr-Cetina 2001, 181)

Epistemic objects are seductive, they always arouse new possibilities of wish fulfillment and at the same time generate a feeling of lack that can be satisfied by the preoccupation with the object. In this sense, epistemic objects are materializations of possibilities that can be realized in a confrontation with the respective object. But these possibilities are never exhausted. On the contrary: Practices with epistemic objects always open new horizons of possibility. Knorr-Cetina typically locates epistemic objects in expert cultures, in which objects are not only used as black boxes in an instrumental or commodified way, but are actively *explored.* Epistemic objects are unfinished entities that generate desires and wishes, full of suggestions for further unfolding.

> [O]bjects of knowledge appear to have the capacity to unfold indefinitely. They are more like open drawers filled with folders extending indefinitely into the depth of a dark closet. Since epistemic objects are always in the process of being materially defined, they continually acquire new properties and change the ones they have. But this also means that objects of knowledge can never be fully attained, that they are, if you wish, never quite themselves. [...] The *lack in completeness of being is crucial*: Objects of knowledge [...] have material instantiations, but they must simultaneously be conceived of as unfolding structures of absences: as things that continually 'explode' and 'mutate' into something else, and that are as much defined by what they are not (but will, at some point have become) than by what they are [...]. The unfolding ontology of objects foregrounds the temporal structure.
>
> (Knorr-Cetina 2001, 181–82)

According to Knorr-Cetina, epistemic objects are essential for the epistemic practices of science and engineering (Knorr-Cetina 1997; Knorr-Cetina 2001). They are indicators of technoscientific "epistemic cultures"

(Knorr-Cetina 1999). One might therefore assume that they only belong into the high-tech laboratories of technoscience. However, as contemporary societies turn into knowledge societies, epistemic objects are becoming parts of everyday life. Hence, according to Knorr-Cetina, we should investigate the "objectual practices" (Knorr-Cetina 2001) with epistemic objects to learn something about the society we are currently living in.

In engineering practices, technical prototypes are conceived as intermediaries between ideas and products. The classical endpoint of a prototyping process is a product ready for mass production. The ultimate success of a product is usually measured and assessed in economic terms: The more products are consumed, the more successful the product. However, the success of the prototype does not depend on such criteria. It can also fulfill his purpose as a unique construction. The prototype is never a ready-to-use product. It is rather a preliminary *projecting* of such a product. According to Alfred Schütz

> projecting consists in an anticipation of future conduct by way of phantasying. [...] Yet projecting is more than mere phantasying. Projecting is motivated phantasying, motivated by the anticipated supervening intention of carrying out the project. The practicability of carrying out the projected action within the imposed frame of reality of the *Lebenswelt* is an essential characteristic of the project. [...] I have to visualize the state of affairs to be brought about by my future action before I can draft the single steps of my future acting from which that state of affairs will result. Metaphorically speaking, I have to have some idea of the structure to be erected before I can draft the blueprint.
>
> (Schütz 1976, 289–90)

Designing something is thus based on the construction of present futures that orient a design practice. According to Schütz, the "blueprints" are aspects of a mental journey into a future. However, prototypes are not purely *imagined* blueprints. Rather, prototypes are materialization of designs. They stand between the design intent and "the state of affairs to be brought about by my future action", in this case: The finished product. Thus, prototypes serve as externalizations and objectifications of projections. They improve the possibilities of representing a future design and they promote communication about the design itself. Through externalization, the design appears as an epistemic object that can be interacted with. Since the prototype illustrates certain functionalities of the desired product, it is itself given the function of acting as an epistemic object to evaluate possible functions of the desired object of the future.

Prototypes, however, do not just fulfill an *epistemic function* but also a social one. They do not only visualize and test a desired future, they also present the future *as* a testable materiality, be it in the R&D laboratory, at a technology fair, or in mass media. The *social function* of prototypes can thus be located in the staging of an invention as possible innovation. Prototypes present projections as something that can be actually experienced. By means of prototyping, the future becomes tangible and gives others an idea of the design goal. The prototype thus appears as a material element of a feedback loop, which in turn reacts to the design process and, if necessary, can offer a new direction. Thus, prototypes are temporally paradoxical objects: They represent a future object (as potential instrument and/or commodity) with which they themselves, as a present (epistemic) object, are not identical.

Compared to other media of futuring, prototypes embody an epistemic primacy of the material over the discursive. Such an attribution of epistemic primacy corresponds with a diagnosis of contemporary knowledge cultures as "technoscientific". Technoscientific cultures of knowledge production valorize a technical usefulness that can be materially demonstrated. Authoritative knowledge in such cultures may be less achieved by carefully developed arguments in academic discourses that aspire to evaluate truth claims, but through demonstrable functionalities (Nordmann 2010).

Epistemic and social function of prototypes are not separate from each other but deeply intertwined. In this way, prototypes are used as media of epistemic *authority* (Gieryn 1983, 1999). A scenario of the future (as a text) may result in disbelief or may be deconstructed as part of a hype discourse. A prototype in contrast, strengthens and solidifies an expectation of a specific future by demonstrating a functioning materiality. In this regard, the relation of a written scenario and prototype is similar to the relation of scientific paper and laboratory infrastructure, as described by Latour. In Latour's "Science in Action" the (fictional) reader in doubt of an author's truth claims is led into to the laboratory. "You doubt what I wrote? Let me show you." (Latour 1987, 64). The demonstration of the laboratory transforms the task of the disbeliever. He now needs to prove that the laboratory instruments are not producing true results or are producing results that do not justify the claims made in the paper. In a similar way, disbelievers in a specific future must either show that the prototype does not perform the function that it claims to perform, or that the function of the prototype does not justify the suggested link between prototypical present and expected future. Thus, the attribution of an epistemic authority of demonstrable objects can be used to enroll other actors in the design of a specific future.

As media of communication, prototypes demonstrate a testable technological future. The prototype promises a future that it cannot currently live up to. Nevertheless, as the prototype tries to assure us, the novel future is not an empty promise, but something that can already be materially tested. Just as research objects are the classical epistemic objects of science, prototypes are focal epistemic objects of practices of design and engineering. No epistemic object symbolizes the contemporary modes of accessing the future as strikingly as the prototype. If objectual practices with epistemic objects are indeed, as Knorr-Cetina suggests, symptomatic for today's expert cultures (Knorr-Cetina 2001), prototyping might be a specific symptomatic practice of the contemporary making of futures. The prototype represents a desire to leave the present behind and to move into a new world – by tentatively creating it.

Designing the Present

This chapter started with a discussion on how scholars may engage with the future. While the future presents remain a realm of the unknowable, present futures are social constructs that can be studied. The recent suggestion for a hermeneutic TA continues academic debates in the social sciences and humanities on such present futures. Hermeneutic approaches typically investigate texts. In this regard, the hermeneutic approach to understand futures is *grosso modo* focused on present futures in the medium of language. It aspires to analyze the content of futuristic discourses as well as the social processes that generate these discourses. This chapter argued that the hermeneutic approach can and should be extended to artefacts as media of present futures and their socio-material production.

This reflects a recent discussion about the production of the future, most prominently articulated by Adam and Groves (2007). They argue that academics have focused a bit too much on specific present futures, but largely ignored how future presents – unknowable as they may be – are constantly produced in the present. To connect research on present futures with this production of the future presents, they suggest that scholars should turn their attention from the analysis of (present) futures as objects of study towards "futures in the making" (Adam and Groves 2007, 37).

Prototypes are an important medium of making *and* communicating the future. Their epistemic function (as externalization of ideas about future technologies that can be materially tested in the present) is entangled with their social function (of demonstrating the plausibility of a present future). However, despite their aspiration to materialize an imagined future and to make it present, prototypes do not overcome existing epistemological boundaries to access *the* future. They still are *present* futures, which do not give access to the future presents of a (potentially) finished technological design.

Prototyping does not only promote certain prototypes; it also promotes the practice of prototyping as a technology of futuring. It valorizes a technology that objectivates futures and makes them "testable" instead of painting a future "merely" by using words and images. Thus, the study of prototypes (as materialized present futures) and prototyping (as a social practice) tells us something about our contemporary technoscientific culture of futurity.

Prototyping is increasingly conceived as a participatory social process that transgresses the borders of engineering labs, design studios, and R&D departments: Heterogeneous groups are invited to take part: From professional innovation experts (engineers, researchers, developers) to creative social communities (artists, makers) to emerging digital collectives (swarms, crowds) (Guggenheim 2010). This inclusion of actors into prototyping and innovation is often facilitated by digital technologies, ranging from crowdsourcing and crowdfunding platforms to low-cost technologies of "digital fabrication" like 3D printers and laser cutters. Thus, prototyping as a socio-material practice of futuring transcends institutional expert cultures of design and engineering and increasingly moves into other social spheres.

Digital (mass) media are an easy way to extend the staging of prototypes beyond the laboratory. Digital publics may not only be assembled as witnesses of prototypical demonstrations; In the case of crowdfunding, they are also assembled as potential donors whose role is to take part in the realization of futures. On crowdfunding platforms like Kickstarter, prototypes are presented to arouse enthusiasm for the final product (Mollick and Robb 2016). Another more active role of publics is an involvement of potential users as beta-testers and co-designers in usability labs or "living labs" (Schuurman et al. 2015).

Moreover, public prototyping workshops like FabLabs and makerspaces have been established in recent years. One important aim of these workshops is to provide an infrastructure to make prototyping more accessible. These workshops are partly institutionalized as companies, partly as non-profit organizations. Digital fabrication technologies like 3D printers and laser cutters make prototyping more easy and affordable. These technologies enable processes of "rapid prototyping" by transforming digital models into physical objects (Buchli 2010).

Another example of public prototyping are civic hackathons. Hackathons are events in which hardware and software developers build prototypes collaboratively within a predefined time frame. In the case of *civic* hackathons, prototyping is performed as a political practice: Citizens are encouraged to develop technical solutions to address local or global issues and challenges. In such events, social and environmental issues are conceived as problems that can be solved by technological fixes (Johnson and Robinson 2014).

As prototyping becomes a normalized medium of future-making we need to scrutinze its message. Just like communicated futures in the medium of language, prototypes are interventions into the present. They transform the present into a domain of tentative designs. If objects are regarded as prototypes, they are framed as unfinished works in progress that ought to be tested and demonstrated. They enter an iterative feedback loop to realize a desired future. Just like the notion of design has been expanded under the paradigm of "design thinking" (Latour 2008; Nordmann 2016; Seitz 2017), the notion of the prototype is currently expanded to less tangible and material phenomena, and even to social institutions (Dickel 2017, 184). Under the technological gaze of prototyping, the present becomes a testbed for the future.

This research was supported by *"Prototyp - Zukunft materiell entwerfen. Prototypen als Kommunikationsmedien des Neuen" (German Federal Ministry of Education and Research, BMBF, 01UO1810A)*

Notes

1 For alternative ontologies of time see for example (Adam 1995; Adam and Groves 2007).
2 Quote translated from German, original emphasis.
3 Indeed, the author of this paper has performed such a hermeneutic analysis of futuristic texts himself (Dickel 2011).
4 For further results of these case studies see (Dickel 2019).

References

Adam, B. 1995. *Timewatch: The Social Analysis of Time*. Cambridge: Polity Press.

Adam, B., and C. Groves. 2007. *Future Matters: Action, Knowledge, Ethics*. Leiden, Boston: Brill.

Agamben, G. 2009. "What Is an Apparatus?" In *What Is an Apparatus? and Other Essays*, 1–24. Meridian, crossing aesthetics. Stanford, Calif: Stanford University Press.

Balka, E. 2000. "Rethinking 'The Medium Is the Message': Agency and Technology in McLuhan's Writings." *Media International Australia incorporating Culture and Policy* 94 (1): 73–87. 10.1177/1329878X0009400108

Barad, K. 2003. "Posthumanist Performativity: Toward an Understanding of How Matter Comes to Matter." *Journal of Women in Culture and Society* 28 (3): 801–831.

Baird, D. 2004. *Thing Knowledge. A Philosophy of Scientific Instruments*. Berkeley: University of California Press.

Bell, W., ed. 1971. *The Sociology of the Future. Theory, Cases, and Annotated Bibliography*. Publications of the Russell Sage Foundation. New York: Russell Sage Found.

Berger, P. L. and T. Luckmann, 1967. *The Social Construction of Reality: A Treatise in the Sociology of Knowledge*.New York: Anchor Books.

Braidotti, R. 2013. *The Posthuman*. Cambridge: Polity Press.

Brown, N., and M. Michael. 2003. "A Sociology of Expectations: Retrospecting Prospects and Prospecting Retrospects." *Technology Analysis & Strategic Management* 15 (1): 3–18.

Buchli, V. 2010. "The Prototype: Presencing the Immaterial." *Visual Communication* 9 (3): 273–286. 10.1177/1470357210372718

Cornish, E. 2005. *Futuring: The Exploration of the Future*. Bethesda, Md: World Future Society.

Dickel, S. 2011. *Enhancement-Utopien: Soziologische Analysen zur Konstruktion Des Neuen Menschen*. Baden-Baden: Nomos.

Dickel, S. 2017. "Irritierende Objekte: Wie Zukunft prototypisch erschlossen Wird." *Behemoth* 10 (1): 171–190. 10.6094/behemoth.2017.10.1.950

Dickel, S. 2019. *Prototyping Society: Zur Vorauseilenden Technologisierung Der Zukunft*. Bielefeld: Transcript.

Gieryn, T. F. 1983. "Boundary-Work and the Demarcation of Science from Non-Science: Strains and Interests in Professional Ideologies of Scientists." *American Sociological Review* 48 (6): 781–795.

Gieryn, T. F. 1999. *Cultural Boundaries of Science: Credibility on the Line*. Chicago: University of Chicago Press.

Gengnagel, C., Nagy, E., and Stark, R. 2016. Introduction. In *Rethink! Prototyping. Transdisciplinary Concepts of Prototyping*, edited by Christoph Gengnagel, Emilia Nagy and Rainer Stark, 1–8. Cham: Springer.

Grunwald, A. 2006. "Nanotechnologie Als Chiffre Der Zukunft." In *Nanotechnologien im Kontext: Philosophische ethische und gesellschaftliche Perspektiven*, edited by Alfred Nordmann, Joachim Schummer and Astrid Schwarz, 49–80. Berlin: Akademische Verlagsgesellschaft.

Grunwald, A. 2014. "The Hermeneutic Side of Responsible Research and Innovation." *Journal of Responsible Innovation* 1 (3): 274–291. 10.1080/23299460.2014.968437

Guggenheim, M. 2010. "The Long History of Prototypes." *Limn* 1.

Gumbrecht, H. U. 2001. "How Is Our Future Contingent? Reading Luhmann against Luhmann." *Theory, Culture & Society* 18 (1): 18–59.

Hölscher, L. 1999. *Die Entdeckung Der Zukunft*. Frankfurt am Main: Fischer.

Hörl, E., ed. 2011. *Die technologische Bedingung: Beiträge zur Beschreibung Der technischen Welt*. Berlin: Suhrkamp.

Johnson, P., and P. Robinson. 2014. "Civic Hackathons: Innovation, Procurement, or Civic Engagement?" *Review of Policy Research* 31 (4): 349–357.

Kaiser, M. 2015. *Über Folgen: Technische Zukunft und Politische Gegenwart*. 1. Aufl. Velbrück Wissenschaft. Weilerswist: Velbrück.

Kittler, F. A. 2006. *Gramophone, Film, Typewriter*. Writing science. Standford, CA: Stanford University Press.

Knorr-Cetina, K. 1997. "Sociality with Objects. Social Relations in Postsocial Knowledge Societies." *Theory, Culture & Society* 14 (4): 1–30.

Knorr-Cetina, K. 1999. *Epistemic Cultures: How the Sciences Make Knowledge*. Cambridge, MA: Harvard University Press.

Knorr-Cetina, K. 2001. "Objectual Practice." In *The Practice Turn in Contemporary Theory*, edited by Eike v. Savigny, Karin Knorr-Cetina and Theodore R. Schatzki, 175–188. London, New York: Routledge.

Koselleck, R. 1979. *Vergangene Zukunft: Zur Semantik geschichtlicher Zeiten*. 1. Aufl. Theorie. Frankfurt am Main: Suhrkamp.

Latour, B. 1987. *Science in Action: How to Follow Scientists and Engineers through Society*. Cambridge, MA: Harvard University Press.

Latour, B. 2008. "A Cautious Prometheus? A Few Steps toward a Philosophy of Design (with Special Attention to Peter Sloterdijk)." Networks of Design, Falmouth, Cornwall, September 3. Accessed April 20, 2018. http://www.bruno-latour.fr/sites/default/files/112-DESIGN-CORNWALL-GB.pdf

Lemke, T. 2015. "New Materialisms: Foucault and the 'Government of Things'." *Theory, Culture & Society* 32 (4): 3–25. 10.1177/0263276413519340

Lösch, A., K. Böhle, C. Coenen, P. Dobroc, A. Ferrari, R. Heil, D. Hommerich, M. Sand, C. Schneider, S. Aykut, D. Fuchs, S. Dickel, B. Gransche, A. Grunwald, A. Hausstein, K. Kastenhofer, K. Konrad, A. Nordmann, P. Schaper-Rinkel, D. Scheer, I. Schulz-Schaeffer, H. Torgersen, and A. Wentland. 2019. "Technology Assessment of Socio-Technical Futures – A Discussion Paper." In *Socio-Technical Futures Shaping the Present. Empirical Examples and Analytical Challenges*, edited by A. Lösch, A. Grunwald, M. Meister and I. Schulz-Schaeffer, 285–308. Wiesbaden: Springer VS.

Luhmann, N. 1976. "The Future Cannot Begin: Temporal Structures in Modern Society." *Social Research* 43 (1): 130–152.

Luhmann, N. 1997. *Die Gesellschaft Der Gesellschaft*. 2 vols. Frankfurt am Main: Suhrkamp.

McLuhan, M. (1964) 2003. *Understanding Media: The Extensions of Man*. Critical ed. Corte Madera, CA: Gingko Press.

Mollick, E., and A. Robb. 2016. "Democratizing Innovation and Capital Access: The Role of Crowdfunding." *California Management Review* 58 (2): 72–87. 10.1525/cmr.2016.58.2.72.

Nordmann, A. 2010. "A Forensics of Wishing: Technology Assessment in the Age of Technoscience." *Poiesis Prax* 7(1-2): 5–15.

Nordmann, A. 2014. "Responsible Innovation, the Art and Craft of Anticipation." *Journal of Responsible Innovation* 1 (1): 87–98. 10.1080/23299460.2014.882064.

Nordmann, A. 2016. "Designs on Design." In *The Intrinsic Logic of Design*, edited by G. M. Buurman and M. Rölli, 8–117. Zürich: Niggli.

Okabe, D., and M. Ito. 2006. "Technosocial Situations: Emergent Structurings of Mobile Email Use." In *Personal, Portable, Pedestrian: Mobile Phones in Japanese Life*, edited by M. Ito, D. Okabe and M. Matsuda, 257–273. Cambridge, MA: MIT Press.

Rheinberger, H.-J. 1992. "Experiment, Difference, and Writing: I. Tracing Protein Synthesis." *Studies In History and Philosophy oF Science Part A* 23 (2): 305–331.

Rip, A. 2009. "Technology as Prospective Ontology." *Synthese* 168 (3): 405–422. 10.1007/s11229-008-9449-9

Schulz-Schaeffer, I., and M. Meister. 2017. "Laboratory Settings as Built Anticipations – Prototype Scenarios as Negotiation Arenas between the Present and Imagined Futures." *Journal of Responsible Innovation* 9 (1): 1–20. 10.1080/23299460.2017.1326260

Schütz, A. 1976. *Collected Papers II: Studies in Social Theory*. The Hague: Martinus.

Schuurman, D., L. de Marez, and P. Ballon. 2015. "Living Labs: A Systematic Literature Review." Open Living Lab Days 2015, Istanbul, Turkei, August 24–August 28.

Seitz, T. 2017. *Design Thinking und Der neue Geist Des Kapitalismus: Soziologische Betrachtungen einer Innovationskultur*. Bielefeld: Transcript.

Selin, C. 2008. "The Sociology of the Future: Tracing Stories of Technology and Time." *Sociology Compass* 2 (6): 1878–1895. Accessed February 3, 2010.

Star, S. L. 1999. "The Ethnography of Infrastructure." *American Behavioral Scientist* 43 (3): 377–391. 10.1177/00027649921955326

Suchman, L., R. Trigg, and J. Blomberg. 2002. "Working Artefacts: Ethnomethods of the Prototype." *The British Journal of Sociology* 53 (2): 163–179. 10.1080/00071310220133287

Theall, D., and J. Theall. 1989. "Marshall McLuhan and James Joyce: Beyond Media." *Canadian Journal of Communication* 14 (4). 10.22230/cjc.1989v14n4a531

Tripathi, A. K. 2016. "The Significance of Digital Hermeneutics for the Philosophy of Technology." In *Information Cultures in the Digital Age: A Festschrift in Honor of Rafael Capurro*, edited by M. Kelly and J. Bielby, 143–157. Wiesbaden: Springer Fachmedien Wiesbaden.

Vetter, M. 2011. *Praktiken Des Prototyping im Innovationsprozess von Start-Up-Unternehmen*. Wiesbaden: Springer/Gabler, Wiesbaden.

Wilkie, A. 2010. Prototypes in Design. Materializing Futures. Limn 1.

10 The Hermeneutic Perspective on Modeling in Technology Assessment

Armin Grunwald

Introduction

Modern societies orientate themselves by anticipating future developments and reflecting the anticipations with respect to desired and undesired aspects, costs and benefits, fears and expectations, chances and risks, etc. (e.g. Luhmann 1992). By reflection and assessment of anticipations, opinions for the future are formed and decisions taken, usually involving uncertainty and risk (Beck 1992). Exploiting this approach needs capabilities for creating, analyzing, and evaluating possible, probable, desired and undesired, plausible, or feared futures. Since the 1950s, the sciences have been developing several forecasting and foresight approaches, methods, and tools for serving the need for anticipation at different levels. Model-based approaches have been developing into a favorite and widespread means for integrating empirical data and scientific insight for providing knowledge-based pictures of the future. The IPCC climate change scenarios (IPCC – International Panel on Climate Change 2014; cp. Poznic and Hillerbrand 2021 for a reflective view), economic growth predictions, energy scenarios for policy advice (acatech 2016), and assessments of the stability of social assurance systems facing demographic change strongly rely on complex mathematical and data based models, often run by high-performance computers. Thereby, model-based anticipatory reasoning has gained a high influence on political decision-making as well as on public debate.[1]

Technology assessment (TA) aims at investigating the consequences of today's decisions with regard to new technology for providing orientation and advice to society and policymakers (cp. Grunwald 2019). TA, therefore, has to work in an anticipatory manner. Data based models play at least a double role. First, TA develops their models, e.g. for conducting prospective life cycle assessments (LCA). Second, TA has to cope with a manifold of already existing model-based futures in many of its fields of research. Often, these are heavily contested (Brown et al. 2000). In both directions, models are used to establish temporal relations: Based on data from the past and knowledge from the present, they shall orientate shaping society's future. Because these temporal

DOI: 10.4324/9781003322290-13

relations include several non-trivial assumptions and expectations leading to the controversies mentioned above, TA is in need of understanding better the roots and origins of the diversity and divergence of the images of the future debated in a certain field, in order to be able to provide reflected orientation to policymakers and society. In-depth consideration of the models and their background is required to this end, at least as far as models had served as the basis for creating the futures under consideration.

A second reason for TA engaging in understanding models is the frequent mistrust in public debate against model-based reasoning, in particular against its exploitation for creating orientation for political measures. Often, in particular, in public debates, the model approach is suspected technocratic, or at least to be at risk of technocracy. As an example, during the Covid-19 pandemic, the suspicion was repeatedly expressed that political measures were mainly driven by models and modelers rather than by democratic political reasoning informed but not determined by model-based insight.

For these reasons, the aim of this chapter is to consider closely the role and epistemological nature of model-based reasoning for creating orientation concerning the future. The field of TA will serve as a case, assuming, however, that similar arguments hold for other types of model-based and future-oriented scientific policy advice, too. Approaching this aim requires referring to temporal structures of modeling. This perspective opens up the door to applying a hermeneutic view on modeling (Erdbeer 2021): Turning the perspective from regarding model-based futures as knowledge about times to come to consider them as contemporary objects to be studied with respect to their content, premises, cultural background, inherent narratives, etc. It corresponds to the hermeneutic turn of TA (and RRI, cp. Grunwald 2014). In this perspective, model-based stories of the future are regarded as expressions of the time in which they have been created, based on modeling techniques, assumptions and narratives, data, and other ingredients of that time. While the hermeneutic turn was developed at the occasion of and often speculative, diverging techno-visionary futures in NEST fields lacking a clear database (e.g. van der Burg 2014; Grunwald 2014), this approach will be extended to the field of model-based futures in this chapter. Here, specific questions emerge, e.g. for models' and modelers' power for influencing decision-makers and shaping the future of society, for the transparency of the models and the 'philosophies' behind, for the narratives behind the models and their exploitation for creating pictures of the future, for the source and quality of the data integrated, and so forth.

The chapter will proceed in the following steps. In the next section, first the use of model-based future reasoning in modern society shall be considered briefly for underpinning the point of departure of this chapter. Subsequently, at least a few ideas concerning the use of model-based reasoning in TA shall be presented, and complemented by referring to the involved circular epistemology. The following section addresses the asymmetric temporality of modeling by reflecting on the circular epistemology in more detail and by addressing the role of narratives in modeling, resulting in the identification of the 'model fix' approach analogously to a techno-fix one. The final section bundles the results gained with respect to reflecting the risk of closing down the future by this 'model fix' and to discussing possibilities for making use of models without running into this trap.

Using Models for Creating and Assessing Futures

For decades, modeling has been developing into a major approach in the sciences for various purposes one of them being anticipating future developments based on current insight and data. Technology assessment applies models to this end, too, but goes beyond in case of diverging future expectations due to different models.

Models for Orientating Decision-Making

Modeling has been a scientific approach to explanation and understanding for centuries, e.g. in astronomy and geodesy. In addition to the property of models of being models *of something*, representing certain attributes of the modeled object (e.g. the solar system or planet Earth) and providing insight into complex relations, models also are used as models *for something*, i.e. for serving a certain purpose as means to an end. For example, models of the solar system served for navigating purposes in antique times already. The duality of *models of* and *models for* (Janich 2002; Nick et al. 2019), resp. the often non-trivial step from using a *model of* in an instrumental manner of a *model for* demarcates a structure significant for the hermeneutic view on modeling.

The availability of computers with strongly increasing capability of collecting, storing, and evaluating huge amounts of data since the 1950s fueled a strong expansion of the modeling approach in science and engineering. The combination of huge mathematical models based on data collections and fast algorithms operating on these models, e.g. for pattern recognition and optimization, allowed integrating and assessing knowledge and data to an unprecedented extent. Since the rise of the cybernetic approach (Simon 1957), models have been used for predicting societal developments, underpinning scenarios of possible future developments, simulating various future worlds depending on certain assumptions, and anticipating new technologies and their impacts (Kahn and Wiener 1967). In particular, models help structuring possible futures in 'what will be?' or 'what if?' statements, often based on 'if/then' relations. Model-based approaches promise opening up new opportunities to better understand, forecast, and control complex socio-technical systems in a data based (and presumed objective) manner by integrating a huge amount of data and cause/effect-relations. The US-American futurologist Hermann Kahn probably was the most prominent person symbolizing almost unlimited optimism with regard to better and better anticipation of almost everything in the course of increasing computer power, enhanced mathematical models, and more and more data.

However, the cybernetic expectations of the 1960s with regard to predicting societal developments and human behavior repeatedly failed. As consequence, the use of models changed in many fields from making predictions to exploiting *models of* for developing scenarios as possible, plausible, or probable futures. As an early example, world models were used in the famous study 'Limits to Growth' (Meadows et al. 1972) for developing scenarios of the World's status in the year 2000, depending on various assumptions. This study did not aim at making predictions but related the future development at the planetary level to human action and behavior in the coming decades. In this approach, the underlying models *of* the world were used as models *for* orientating human action, intending creating awareness and motivating change of consciousness, behavior, and policies in order to prevent disastrous developments. In this sense, the study followed a pedagogical approach using if/then-statements. In more recent times a revival of the early cybernetic expectations concerning the possibility of prediction can be observed. Strongly increasing computing power, fast algorithms equipped with artificial intelligence (AI), and huge amounts of data collected by a strongly increasing number of ubiquitous sensors and by Internet transactions motivate enthusiasts to expect full predictability of future life. That community shares a strong belief in the calculability of the future as soon as enough data and AI capability will be available.[2]

Among the major current application fields of model-based future reasoning are climate change and its consequences for society and the environment, energy (Aykut 2019), the management of scarce geo-resources, health issues, social assurance systems, and the future of the labor market. The field of energy in combination with climate change and the need for reducing greenhouse gases is among the most prominent areas of applying complex computer models for exploring future developments in order to create policy advice for political institutions but also to companies and managers for orientating decisions on investment:

> With a view to the multitude of factors determining the energy system, most scenarios are created by means of computer models. The variables of the model reflect the characteristics of the system and the most important influencing factors. Such features are, for instance, the installation and operating costs of different technologies for energy generation and use, their technical efficiency rates, or the amount of carbon dioxide (CO_2) emissions they account for. To compute a scenario, numerical assumptions will have to be made, i.e.

> specific figures determined, for all variables not calculated by the model. This refers to e.g. fuel prices, the development of the energy demand or the development of the efficiency rates of individual technologies. By determining specific values for each of these parameters, one scenario is selected from the range of possible ones. Based on that, the values of the other variables are computed, for instance the total amount of carbon emissions or the overall cost of energy generation.
>
> (acatech 2016, 9)

The computer models mentioned – and this holds for models in other fields, too – claim to be models *of the energy system* or of parts of it. They shall represent specific properties and attributes of elements of the modeled systems, of their internal relations as well as of their import/export relations with their social and natural environment. The validation of the models happens, ideally, by comparing data from the real world to the modeled world. After validation in this respect, the models are applied for developing simulations, scenarios, and predictions, which, in the next step of consequentialist reasoning (Darwall 2002) are to be exploited for gaining some orientation for decision-making and action.

Models and Consequentialist Reasoning in Technology Assessment

Technology Assessment (TA) aims at creating and assessing prospective knowledge about possible, plausible, or probable *consequences* of technology, be they intended or unintended, for drawing conclusions for orientation and decision-making (Grunwald 2019; cp. Figure 10.1 below). This usually happens in the *consequentialist paradigm* (Darwall 2002), e.g.

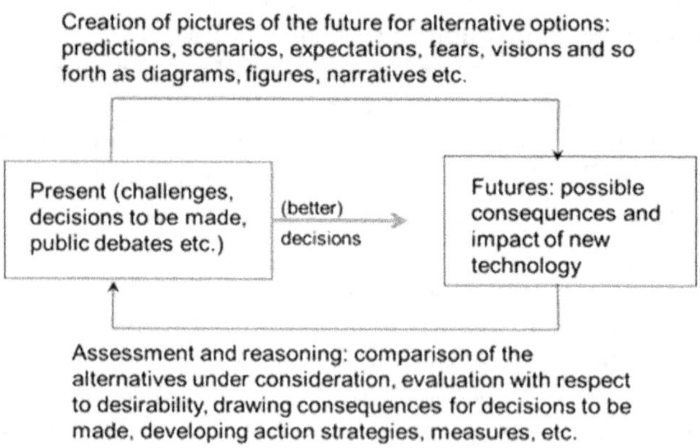

Figure 10.1 The consequentialist paradigm of TA.

Source: Grunwald 2012, 313, modified.

as early warning against risks or early discovery of benefits. In consequentialism, the orientation for decision-making and action is to be derived from assessing the possible consequences of an action with regard to certain criteria such as costs and benefits, desirability and ethical compatibility. Therefore, the availability of validated knowledge about future consequences (Bechtold et al. 2017) is the major premise of consequentialist approaches in consequentialist reasoning in general and in TA, in particular (Figure 10.1).

Frequently, models of technical or socio-technical systems of the present (e.g. the energy system mentioned above) are used for gaining validated extrapolations to the future, depending on certain assumptions and exposing those systems to possible action strategies. Beyond extrapolation, the modeling approach also offers a validation strategy for giving evidence to the consequentialist loop (Figure 10.1): The evidence of the future statements needed for providing orientation is borrowed from the quality of the underlying *model of,* which usually is ensured by empirical data and insight. Insights gained in the *model for* perspective are legitimated by referring to the validity of the *model of.* While model-based future statements cannot be validated against independent data of the future, models serve *as a substitution* for lacking independent validation. In this sense, *models of* are a 'crutch': They shall compensate the lack of empirical validation of the future statements by hinting to their own empirical basis taken from the past. Ignoring the nature of this 'crutch' opens up risks of fixing and closing down the future bound to the respective model.

Data based models play a double role in TA. On the one hand, TA develops models for creating validated images of the future in a certain respect, e.g. for conducting prospective life cycle assessments (LCA) to learn about the sustainability impacts of new technology. On the other, in TA has to cope with a manifold of images of the future already existing in many fields of analysis, many of them rooted in data based models such as energy scenarios (acatech 2016). Usually, the respective future expectations are contested (Brown et al. 2000) and wave, partially, between paradise-like expectations and dystopian fears (Grunwald 2007). While this situation is almost obvious in the field of techno-visionary futures (Brown et al. 2000), e.g. for human enhancement visions (e.g. Nordmann 2007), it also applies to model-based futures, e.g. energy scenarios (Grunwald 2011).

Regarding the necessity of valid prospective knowledge (Bechtold et al. 2017) in the consequentialist scheme of TA (and beyond), the diversity and often divergence of images of the future threaten the feasibility of consequentialist ethics (Nordmann 2007) and TA (Grunwald 2014). In the worst case, i.e. in the absence of valid prospective knowledge and completely diverging assumptions of the future, the consequentialist loop (Figure 10.1) would break down. Any conclusion drawn in this loop would become arbitrary or ideological. The entire loop would degenerate into an epistemologically void, vicious circle.

This worst case, however, is only one endpoint of a continuum. In a strict epistemological sense, even conclusions from the present time to the future are circular, in a particular sense, because they are bound to the immanence of the present.

> But even now it is manifest and clear that there are neither times future nor times past. Thus it is not properly said that there are three times, past, present, and future. Perhaps it might be said rightly that there are three times: a time present of things past; a time present of things present; and a time present of things future ... The time present of things past is memory; the time present of things present is direct experience; the time present of things future is expectation.
>
> (Augustine 397, XI, 20)

Future statements are just expectations based on current knowledge and data from the past but also composing several ingredients such as values (Diekmann and Peterson 2013), assumptions, and interests, all of them belonging to the present time (Luhmann 1976), where, according to Augustine, direct experience and action is possible. Therefore, the epistemological validity of any conclusion for the future is restricted to the present time. This diagnosis does not render valid future statements impossible. On the contrary, in many fields like astronomy, the prediction has a great tradition and works to an extremely high extent. Nevertheless, also predictions in astronomy remain – well-founded – *expectations* because no data of the future are available. In many other fields, in particular, with respect to futures of social developments, the evidence for making the prediction is much lower. Because the futures to be considered in TA result from interrelations between new technology and social developments such as human behavior, political regulation, and values, the latter also holds for TA (cp. also Nordmann 2014). In several cases, the evidence of these futures is questionable or even may disappear, as mentioned above. Hence, while circular epistemology applies to any future statement in the immanence of the present mentioned above their evidence may vary from strong to almost vanishing, depending on the respective epistemic situation. Among the functions of models is to enhance the evidence of future statements by following the 'crutch' metaphor mentioned above.

In case of low or even nonexistent evidence of anticipatory statements, the validity of consequentialist arguments disappears. In those cases – many examples have been uncovered in the field of NEST developments (e.g. Nordmann 2007; Grunwald 2007) the consequentialist loop (Figure 10.1) breaks down epistemologically but would degenerate to an epistemologically void *circulus vitiosus.* In this case, any conclusion taken in spite of this situation would run into the problems of arbitrariness or ideology generally associated with circular situations. It would open the door for interest-driven approaches (Beckert 2016) and self-fulfilling developments (Merton 1948; see Sect. *Narratives in modeling* for examples in the context of modeling). Therefore, the applicability of consequentialist reasoning depends on the respective case and must be checked for each TA study according to the evidence of the prospective knowledge available. In case this approach is insufficient or completely inapplicable, other modes of orientation are required.

Then, a *hermeneutic perspective* must complement the consequentialist approach of TA (cp. Grunwald 2014). While stories of the future are regarded as information about what can be expected to come under certain conditions in the consequential perspective, the hermeneutic perspective takes these stories as expressions of *present time's* diagnoses, attitudes, observations, values, and interests. Thus, a certain future statement, e.g. a model-based energy scenario or a narrative about societal futures with AI, can be analyzed in both the consequential and hermeneutic perspective, with different but complementary cognitive interests behind, with different methods, and with different types of results. In hermeneutic respect, stories of the future are regarded as documents of today's debates. Their analysis shall allow uncovering backgrounds, presuppositions, and pre-occupations with the primary objective to clarify the *contemporary meaning* of the statements and narratives of the future in the respective TA context (Grunwald 2019). In this respect, underlying models as essential ingredients for creating prospective knowledge, e.g. energy scenarios, become the subject of study from the hermeneutic perspective.

Hermeneutics of Model-Based Futures

Key words from the preceding section are the distinction model of/model for, the immanence of the present, circular epistemology of future statements, and hermeneutic approach. In this section, I will dig deeper into the asymmetric temporality of using models for providing future statements of the future and its possible problematic implications. These hold, in particular, for *narratives* underlying and influencing the process of model-building.

Asymmetric Temporality of Modeling

Modeling is, generally speaking, the methodologically controlled composition of ingredients such as data, knowledge about causal or statistical relations, system boundaries, assessments of relevance, values, and interests, pre-determined values of particular parameters, e.g. as assumed megatrends, ceteris paribus assumptions, mathematical representations, background theories e.g. from macro-economics, assumptions about human behavior, and so forth.

Models serve as means for improving *understanding* of the modeled system in the respective present time (*models of*) as well as means for supporting future-oriented reasoning towards *decision-making and intervention* (*models for*). Hence, models are, metaphorically speaking, twins of the modeled systems *in certain respect*. This *certain respect* is of eminent significance for understanding the nature models. It is determined by the purpose the *model for* shall serve. It allows, for example, distinguishing between relevant and less relevant parts of the system and, subsequently, sorting out less relevant factors for building the model. Hence, the representative perspective of the *model of* is inherently connected to the instrumental perspective of the *model for* (Janich 2002; Nick et al. 2019). With specific attention to the issue of future being the focus of this chapter, a similar observation was published recently at the occasion of an analysis of energy modeling:

> When used in practical problem-solving contexts, models integrate both epistemic and non-epistemic values, since applied models are shaped by specific problem framings. The overall quality of these models does not, therefore, rely only on epistemic criteria, but also on pragmatic solution-oriented considerations and on desired behaviours of the target system. This raises important and often overlooked questions on how epistemic values are embedded alongside non-epistemic values in the modelling process.
>
> (Silvast et al. 2020, 1)

Models *of something* shall assemble all the relevant knowledge available required *for answering certain questions* in TA and beyond. Models must be designed, created, filled with data, validated and exploited according to the cognitive interest behind. Models have authors, individual researchers but also huge teams in scientific institutes, companies, and consulting agencies. The creation of models can be regarded analogously to the production of scientific knowledge in a lab. It consists of a social process with actors making decisions about model design, boundaries of the modeled system, distinctions between relevant and less relevant data and relations, validation strategies, and so forth (Dieckhoff 2015). The respective models are created in accordance with available knowledge, assembled usually in the mathematical models but also with reference to assessments of relevance, value judgments, and interests, often in the framework of commissions awarded by decision makers in politics and the economy (acatech 2016).

As any human action, modeling and applying models for drawing conclusions happen over time. Models have to be developed, and this happens, obviously, at a present time. Their construction needs time and uses ingredients such as knowledge, values, and visions belonging to the inventory of that time including its past. While data from the past are available (such as documents, ruins, or images) data from the future are not. While statements of the past can be validated by referring to empirically observable relicts, this is not possible for statements of the future. Modelers creating predictions, scenarios and simulations cannot break out of the present. They only can provide future statements based on present knowledge, data, values, and narratives.

Reviewing the report 'Limits to Growth' (Meadows et al. 1972) today with respect to the models used, we can learn about model techniques, narratives, and assumptions of the time in which it was written. While models are made by using ingredients and processes of a certain time, they may have an impact far beyond in the future – this is the *temporal asymmetry* of modeling. Independent from their epistemic validity, futures, including model-based ones, are disseminated by individual authors, teams, journalists, scientists, policymakers, science managers, and others. They are communicated via different channels, journals, networks, mass media, research applications, expert groups, policy advice, etc. (Selin 2008). Different utensils and formats are used to communicate and disseminate futures: Texts of different kind, diagrams, graphs, pictures but also pieces of art and movies (Grunwald 2014).[3] Some futures will motivate actors and groups to create real-world impacts, e.g. by influencing public attitudes towards new technology, or by convincing policymakers or funding agencies to support and promote the new ideas by public funds.

For example, the report to the Club of Rome (Meadows et al. 1972) was highly influential in creating awareness with respect to the limited availability of resources on planet Earth with societal and political consequences until today. It still has some influence today in spite of the fact that the underlying models and data are more than fifty years old and far from up to date, and had received heavy methodological criticisms already in the 1970s. By doing this, the initial model *of the world* is used as a model *for realizing certain purposes* and serving certain interests for creating orientation and shaping some future developments. The duality of model *of* and model *for* (Janich 2002) transforms into an asymmetric temporal relation: While models are used in TA *for* anticipating future developments, they remain as models *of something* bound to data, insights, values, and other ingredients *of the past and the present.* A deep-ranging asymmetry over time heavily affects the use of models in this respect and is at the core of related challenges: While no data of the future are available for *validating* the models, these become decisive for *shaping* it. *Models of* already can even predetermine and fix at least some aspects of a *model for* which implies predetermining future developments based on the respective model (*model-fix*).

A major observation of this chapter is the concurrence of two insights, each of them already alone implying heavy challenges: The temporal asymmetry of modeling unfolded above and the circular epistemology of future reasoning, in particular of TA. These both meet on the occasion of model-based futures used in TA raising the question of risks of biased or arbitrary conclusions because circular reasoning and asymmetric temporality open many doors for interest-driven futures, self-fulfilling stories (Merton 1948), the exertion of power due to influential narratives of the future (Sand 2019), and so on. They also raise questions with respect to the openness of the future versus its fixation by model-based reasoning (model fix, cp. Sect. *Model-fix: Fixation of the future by underlying models*) and the respective exertion of power over the future by modelers being neither willing nor legitimated nor mandated to do this.

Narratives in Modeling

Repeatedly it was stated (e.g. Walton 1990; Toon 2014; Roßmann 2020) that models, including mathematical and data based ones, are more than the representation of data composed according to empirically validated facts (*Modellüberschuss* according to Mahr 2015). Models involve narratives implicitly or explicitly shared among their authors or in the respective communities involved. These narratives have an orientating role in composing the data available and other ingredients used in modeling and for integrating them into a consistent and coherent picture. They come to the fore when gaps in data, assumptions, and other ingredients must be filled for the sake of theoretical or practical coherence: "Narratives have the function to organize the materials and make objects recognizable as props of a coherent whole" (Roßmann and Rösch 2019). There is no logical necessity or force to build a particular model based on a set of ingredients but rather, usually there will be several alternatives for composing the ingredients into a consistent and coherent model. Against this observation, the function of narratives in composing the ingredients is to close gaps and bridge heterogeneous parts of the set of ingredients. They function as a kind of compass to guide the composition process under certain overarching storylines. Consequently, the model-based future statements frequently used in TA are not only data- and knowledge-driven but also may transport traces of the underlying narratives. Therefore, the societal meaning of new technology as the core object of TA investigations (Grunwald 2019) may be related to the narratives involved in modeling. If TA practitioners make use of models for exploring and assessing future statements they have to be aware of the societal meaning perhaps already having been embedded into the models by the respective authors, following their mindsets or even intentions. Therefore, it is crucial in TA to identify the *place* and the *functions* of narratives in modeling as well as their *content*. Typically, this is a hermeneutic task.

Narratives are present in modeling at different levels of abstraction. At a rather general level, the *science knows best* narrative (Pielke 2007), the *value-neutrality* narrative, the *objectivity* narrative, and the *nimbus of data* narrative can be identified (Erdbeer et al. 2021), claiming scientific superiority in a traditional sense and telling a lot about the self-understanding of modellers' communities. More specifically, narratives behind strands of scientific reasoning and theoretical background convictions play a crucial role. Examples are stories claiming the superiority of scientific disciplines (e.g. so-called hard-core sciences over humanities) or representing specific disciplinary traditions, e.g. neoclassical economics and the image of the *homo oeconomicus*. At a more down to Earth level, narratives also may stand behind substantial convictions and belief systems in certain fields. In energy modeling for the transformation of the energy supply system, for example, often predeterminations in favor of specific future solutions can be found, e.g. preferring nuclear power (ecomodernist manifesto 2015) or renewables. Similarly, in the field of mobility, models *for* often are not technology-open but favor certain lines of development, e.g. e-mobility or hydrogen technologies.

According to the asymmetric temporality of modeling, the narratives involved in these models may have a substantial influence on public debate and decision-making in politics and in the economy. An example is the model-based prediction of the energy demand in Germany and other Western countries in the 1960s. A dominant narrative of that time was assuming an annual GDP growth of 3% per year, related to an annual increase of the energy demand of also 3%. It was a common conviction that economic growth implies the growth of energy consumption in the same magnitude. Based on the resulting model-based predictions of the energy demand power plants were planned and built, and investments were taken. However, following the two oil crises in the 1970s and the growing consciousness of limited resources and global environmental change, policies of enhancing the efficiency of energy use were established. Economic growth was partially decoupled from an increase in energy demand, in contradiction to the underlying narratives of the 1960s. Because many power plants had already been built or were under construction according to the previous predictions, an over-supply of energy in the 1980s and 1990s was the consequence in many countries. This example demonstrates the power of narratives, in particular of *Zeitgeist* issues in modeling. While hermeneutic analysis of those narratives is expected to improve understanding on time for allowing better-reflected decisions this is in particular difficult to be done contemporarily.

An even worse concern is that, in this situation, modelers could heavily influence political and public debates around new technologies or transformations, e.g. of the energy system, *by intention*. The use (or misuse) of narratives used in modeling could be part of an agenda to support certain assignments of meaning and to weaken others. The circular epistemology tolerates this approach – there are no independent data of the future available to prove such attempts false. In its extreme form, self-fulfilling prophecies could be created by driving development over time into the desired direction by self-reinforcing effects. In this way, the approach to the future would not be open to reflecting many alternative options, assessing them, and then coming to a transparent decision, as is the ideal of TA (Grunwald 2019). Instead, first, the endpoint would be fixed as the target, while the process for reaching it would be organized accordingly later, even with possible impact on decisions to be made in modeling in the sense of *model of* and exploiting this in the *model for* perspective. This approach corresponds to technocracy as demarcated by Habermas (1970).

Models can even have self-fulfilling power due to underlying narratives including substantial preoccupations. Imagine, a particular view on mobility aspects of the German *Energiewende* (Aykut 2019) would be shared in a particular community of modelers, e.g. the conviction of an *a priori* superiority of hydrogen technologies over e-vehicles in the future. The modelers would include this view in their models, establish energy scenarios based on those models and convince TA to use them as a source for policy advice. Based on this advice, policymakers would commission further studies, which would support the initial ones because the respective modelers share the same basic views. In this way, the particular view of particular modelers could influence, even determine political decision-making and thus be responsible for a self-fulfilling story. This thought experiment recalls risks of technocracy, hidden, however in the domain of modelers often claiming to be value-neutral, objective, and free of own their interests. The story tells about possibly non-recognized and implicit risks of applying a consequentialist mode of creating orientation beyond its area of validity.

The hermeneutic perspective

Using models in future reasoning is promising but also challenging as results from the preceding section show, mainly because of the required transfer of the *model of* to the *model for* perspective, or: Because of the mixture of both. In particular, the analysis demonstrated not only the significance of narratives in modeling but also their inherent ambivalence: Narratives guide the modeling processes and can lead to unrecognized bias, one-sided developments, and even to ideological applications. Applying the hermeneutic perspective shall contribute to avoiding losing the space of futures open to deliberation in favor of technocratic trust in optimization models.

Model-Fix: Fixation of The Future by Underlying Models

When narratives in modeling (see above) remain implicit and subliminal with non-reflected premises and implications, they could trigger an unrecognized semantic overflow, a 'meaning in disguise', leading to intendedly biased or unintended impacts on the orientation expected. The challenges and problems are rooted in the confusion of the model *of* and the model *for* perspective. In particular, this confusion concerns the claimed *evidence* of model-based reasoning. The evidence of the models applied is usually checked empirically by referring to data available. In exploiting the models for creating prospects to the future, however, this evidence is – again: Usually – transferred to the future statements. The future statements, e.g. predictions of the energy demand, i.e. the results of a *model for* consideration, are regarded evident because the underlying *model of* had been validated according to the state of the art. If done in an unquestioned manner, this approach would fix the future to a model validated by data from the past.

In particular, the concurrence of narratives such as *science knows best*, *objectivity*, and *value-neutrality* culminate in a widespread but (in the TA perspective) dangerous mixture: The *optimization narrative*. According to this narrative, there must be one best solution, which modellers and scientists are promisingly looking for. As soon as they have calculated the optimal solution to a problem, based on data-based and mathematical models, this solution will be recommended to politics and society, claiming that it should be implemented simply because it is regarded as the objectively optimal one. I will call this the *model-fix approach*, analogously to the techno-fix preoccupation. Among the implications of this narrative are:

- the political dimension of making choices according to various values, positions and interests, e.g. in the field of energy transformation, would disappear
- consequently, democratic deliberation would be replaced by a model-based optimization calculus
- instead of exploring alternative pathways to the future, the space of possible futures as an at least somewhat open space would be closed down to the optimized pathway
- the future, hence, would go lost. The underlying model and the data are, as was shown above, epistemologically bound to the past. Applying this model for optimization implies prolonging past relations and trends to the future
- the perspective of shaping the future according to values and desires would become more or less obsolete in favor of adaptation to the implications of the model based pathway selected (this happens analogously to the technology determinism approach (Grunwald 2019)
- the opportunity of shaping the future would be given to modelers with their underlying and dominant narratives, independent of whether they intend to shape the future or not
- in this manner, the future would be fixed according to a mixture of narratives, knowledge and other ingredients of the respective present time, and data from the past

While this list already clearly demonstrates what is at stake, often a second order of fixation can be observed: The reification of models. As soon as models have been built they belong to the interior of science and can be used as existing objects for different purposes. Often, in modelers' colloquial language the phrase is used: "The model tells us", or similar phrases. Beyond reification issues, such phrases tell about – probably unintended – attitudes anthropomorphizing the models. Models then are regarded as a kind of actor. In Actor Network Theory (Latour 2007), they could be regarded as actants with their own agenda. No question about how influential then models can be.

The major risk of the fixation of models as objective and value-neutral representatives of reality is fixing the future due to a specific model with its specific perspectives grounded in the ingredients chosen, the decisions made, and the narratives involved (see the fictive example in sect. *Narratives in modeling*). In the worst (and most devastating) case, the future would be 'colonialized' by the past through modeling, supported by the narratives mentioned, while reasoning and deliberating about alternative options, about the desired and of the future as an open space would be prevented (Stirling 2008). Leaving futures to models, and models to modelers taking their results concerning the future for granted without questioning them, would implicitly imply accepting a kind of involuntary technocracy by modelers. This scenario illustrates that the issue under consideration in this chapter is not an academic one only but of high significance for scientific policy advice, technology assessment, public understanding of science (including scientific modeling), democracy, and political decision-making.

Towards Hermeneutically Informed Inclusive Modeling

As a result, the risk of fixation and reification (and considering the "anthropomorphization" of models as actors) became clear: It is about losing the future or, better, giving the power for shaping the future to models. The hermeneutic view on models and modeling shall, in line with the hermeneutic turn of technology assessment, contribute to opening up alternatives instead of closing down decision-making challenges to a particular solution claimed to be the "best" one (Grunwald 2019; cp. also Sect. *Models and consequentialist reasoning in technology assessment*). Applying the hermeneutic perspective implies first regarding models not as objects *per se* but as *constructed* objects having authors with intentions, values, interests, and so forth. In a hermeneutics of modeling for uncovering possible fixation and for regaining openness, the *process of modeling* as human activity must be considered: Understanding models needs understanding the process of their making (e.g. Diekmann and Peterson 2013; Dieckhoff 2015). During this process, many decisions have to be made at each step, in particular for distinguishing relevant and less relevant aspects while determining system boundaries, for determining the ingredients to be taken into account, and for the choice of the model type including its background convictions and assumptions. Narratives may enter or influence this process at any stage, allowing considering even the entirely resulting models as narratives (Erdbeer et al. 2021). In this sense, data based models are regarded as narratives with embedded quantitative elements. These elements, the data, and their mathematical representation, also have stories behind them: Data have been collected with respect to purposes, based on decisions on what data had been recognized relevant for that purpose, and by applying certain measures of data collection with their own normative theory of measurement behind. The often claimed objectivity and value-neutrality of models (cp. Sect. *Narratives in modeling*), therefore, has to be questioned by hermeneutic analysis.

In this respect, the model theory of Mahr (2004, 2011, 2015) provides a framework in line with the proceeding argumentation on the model of/model for structure and the temporal asymmetry of modeling. According to Mahr, nothing is a model by itself but everything *can be perceived* as a model in a certain context as a model of or a model for (Mahr 2011), preventing any reification of models. Mahr emphasizes that the actualizations of the model's possibilities is largely due to the decisions of modellers exploiting an existing model for particular purposes as a model *for*. The being of a model as well as making use of it to certain ends is regarded as the result of intentional acts.

In particular, Mahr's concept of the *Befragbarkeit* (possibility of addressing questions) of models (Mahr 2015) directly relates to the preceding para. Questions can be raised with regard to the meaning of models, their authors and history, the narratives involved, and their cultural background, to the chosen modeling approach and its normative presuppositions, to the data and their biographies, to underlying assumptions, and so forth. Applying *Befragbarkeit* means raising questions for what reasons the models were built in the particular manner chosen and not in another one. Some important implications are inherently related to this kind of questions:

- the reification of the model under consideration will be broken
- the objectivity often claimed will be relativized by the insight that the model is a result of many decisions in the preceding modeling process made in means/end rationality, either in the model of or in the model for approach
- any claim of closing down the future development considered to an "optimal" pathway would get deconstructed. Because many decisions during the "making of" of the model could also have been made in another way, according to other purposes, also the conclusions for the future would vary

- therefore, applying the concept of *Befragbarkeit* undermines any "there is no alternative" approach (TINA) but allows for reconquering "thinking in alternatives" for shaping the future

In the immanence of the present time, the hermeneutics of potential futures based on models turns into a *hermeneutics of contemporary expectations* (*Modellerwartungen*) that influence or even create technological and social futures. Hence, raising questions about possible alternatives and for their properties and implications is, in this sense, *subversive* with regard to closing down futures, e.g. by optimization models. It allows, however, maintaining the functions of models urgently needed in modern science and scientific policy advice: Integrating huge amounts of data of different origins, considering systemic feedback loops, and exploring the future like in the field of TA. By applying models in this understanding, public deliberation and political decision-making can be informed but will not be replaced, avoiding the technocracy concern (see above). This approach corresponds nicely with a model of scientific policy advice proposed recently (Edenhofer and Kowarsch 2015). It was adapted to the specific field of climate policy, which heavily relies on huge models. In this approach, researchers are seen as mapmakers while policymakers shall use the resulting map as point of departure for determining where the travel shall go:

> … researchers, along with stakeholders, act as the "cartographers" of different, viable policy pathways and their practical consequences by acting as the "mapmakers" of the political solution space. They provide a guidebook with alternative options for policymakers (.i.e. the "navigators" and the public). Such maps cannot replace travelling i.e. decision-making nor can they resolve all environmental policy conflicts, yet they can provide an important orientation in otherwise uncharted territory.
>
> (Edenhofer and Kowarsch 2015, 63)

A second sensitivity of applying models in scientific policy advice – beyond the issue of closing down futures, e.g. by optimization models – can be found in the area of playing around with what is regarded possible or feasible in a certain field. Models are used in policy advice in general and in TA in particular also for allowing making statements about the possibility or feasibility of certain future developments, e.g. in the field of energy (acatech et al. 2016). In this manner, models influence the range of potentialities, possibilities, and opportunities for being debated in the public and in policy-making.[4] A specific model demonstrates particular futures as possible or plausible while pushing other futures out of the range of possibility or plausibility. Thus, the *Befragung* of models with regard to the process of modeling (see above) and the subsequent analysis may help uncover path-dependencies, biases, and fixations in this respect, too.

The hermeneutic view, therefore, means uncovering the particularity of a specific model according to its construction process and context. By recognizing this particularity, the futures based on a specific model can be regarded as possible ones amongst other possible ones based on other models. This step opens up thinking about alternatives as well as the possibility to compare and weigh different options with respect to implications and consequences.

Conclusion

Model hermeneutics is a critical procedure operating on the borderline of data-oriented representation of parts of the real world (models of), making use of those representations e.g. by generating future prospects (models for), both of them involving *social narratives.* It shall elucidate both the authority and the fragility of models in the circular epistemology of future reasoning and add evidence to the assumption that talking more in-depth about models and modeling as a process is essential for better understanding the construction of futures. Seeking improved understanding must, in accordance with the Augustinian view on futures (see above), focus on the *present ingredients* and the processes of modeling instead of looking at their results as projected developments of the future. The recognition of the *immanence of the present* leads to a shift from an anticipatory or even prognostic view on model-based futures, e.g. energy scenarios, to analyzing their current processes of construction, the ingredients used, and the narratives behind. In line with the hermeneutic turn of technology assessment, this shift takes model-based futures as expressions of the present time into consideration and allows re-conquering the future as an (at least partially) open space for thinking in alternatives in contrast to narrowing down this space by applying particular models.

Beyond analytic and reflexive approaches to better understanding modeling and avoiding misinterpretation by fixation and reification, relegitimizing modeling in the eyes of the public and of policy-makers needs establishing a better reflected 'model literacy'. This should exploit the insights from the hermeneutic perspective on modeling for establishing a new view on models and modeling in the modeling communities themselves, e.g. by integrating this perspective in university education and in the modelers' communities.

Acknowledgement

The analysis and conclusions presented in this chapter partially are results of the project Poetic Modeling and Energy Transition funded by the Volkswagen Foundation (cp. Erdbeer 2023).

Notes

1 These capabilities are neither equally distributed over actors and groups of the population, nor do all members of society have equal opportunities to contribute to the influential future debates in societies (cp. the chapters by Chris Groves and Martin Sand in this volume).
2 This optimism ignores epistemological barriers, e.g. the issue of self-fulfilling or self-destroying prophecies (Merton 1948), and human freedom to behave differently in the future than in the past (Schwemmer 1976).
3 The underlying models themselves usually are not communicated, some of them are even kept confidential due to business interests (acatech 2016).
4 This observation is similar to an effect of new technology, which is rarely discussed: itIt not only opens up new opportunities for action but also influences what is regarded technologically possible beyond that technology (Grunwald 2018).

References

acatech - National Academy of Science and Engineering, ed. 2016. *Consulting with Energy Scenarios: Requirements for Scientific Policy Advice* (available at www.acatech.de).

Augustine of Hippo. 397. Confessiones XI, 20.

Aykut, S. 2019. “Reassembling Energy Policy: Models, Forecasts, and Policy Change in Germany and France.” *Sci. Technol. Stud.* 32 (4): 13–35. 10.23987/sts.65324

Bechtold, U., D. Fuchs, and N. Gudowsky. 2017. “Imagining socio-Technical Futures – Challenges and opportunities for Technology Assessment.” *Journal of Responsible Innovation* 4 (2): 85–99.

Beck, U. 1992. *Risk Society: Towards a New Modernity*. New Delhi: Sage Publications. (Original: Risikogesellschaft. Auf dem Weg in eine andere Moderne. Frankfurt a.M.: Suhrkamp.)

Beckert, J. 2016. *Imagined Futures. Fictional Expectations and Capitalist Dynamics*. Cambridge, MA; London, England: Harvard University Press.

Brown, N., B. Rappert, and A. Webster, eds. 2000. *Contested Future: A Sociology of Prospective Techno-Science*. Burlington: Ashgate Publishing.

Darwall, S., ed. 2002. *Consequentialism*. Malden, MA: Wiley.

Dieckhoff, C. 2015. *Modellierte Zukunft - Energieszenarien in der wissenschaftlichen Politikberatung*. Bielefeld: transcript.

Diekmann, S., and M. Peterson. 2013. “The Role of non-epistemic Values in Engineering Models.” *Sci. Eng. Ethics* 19 (1): 207–218. 10.1007/s11948-011.

Edenhofer, O., and M. Kowarsch. 2015. “Cartography of Pathways: A New Model For Environmental Policy Assessments.” *Environmental Science and Policy* 51: 56–64.

Erdbeer, M., E. Achermann, A. Grunwald, V. Hagenmeyer, and K. Stierstorfer. 2023. *A Critique of Modelling Judgment*. London: Routledge (to appear).

Grunwald, A. 2007. “Converging Technologies: Visions, increased contingencies of The Conditio Humana, and Search for Orientation.” *Futures* 39 (4): 380–392.

Grunwald, A. 2011. “Energy Futures: Diversity and the Need for Assessment.” *Futures* 43 (8): 820–830.

Grunwald, A. 2012. *Technikzukünfte Als Medium von Zukunftsdebatten und Technikgestaltung*. KIT Scientific Publishing: Karlsruhe.

Grunwald, A. 2014. "The Hermeneutic Side of Responsible Research and Innovation." *Journal of Responsible Innovation* 1 (3): 274–291.

Grunwald, A. 2018. "Technik als Transformation von Möglichkeitsräumen: Technikphilosophie anders gedacht." In *Möglichkeiten der Reflexion: Festschrift für Christoph Hubig*, edited by P. Richter, 203–216. Baden-Baden: Nomos.

Grunwald, A. 2019. *Technology Assessment. Practice and Theory*. Abingdon: Routledge.

Habermas, J. 1970. *Toward a Rational Society*. London: Heinemann. (Original: 1968. Technik und Wissenschaft als Ideologie.Frankfurt a. M: Suhrkamp.)

IPCC – International Panel on Climate Change 2014. "Fifth Assessment Report." Assessed July 11, 2021. www.ipcc.ch/report/ar5/syr/.

Janich, P. 2002. "Modelle und Modelliertes." In *Integrative Modellierung zum Globalen Wandel*, edited by C.F. Gethmann and S. Linger, 48–70. Berlin: Springer.

Kahn, H., and N. Wiener. 1967. *The Year 2000: A Framework for Speculation on the Next Thirty-Three Years*. New York: Macmillan.

Latour, B. 2007. *Eine neue Soziologie für eine neue Gesellschaft: Einführung in die Akteur-Netzwerk-Theorie*. Frankfurt a.M.: Suhrkamp.

Luhmann, N. 1976. "The Future Cannot Begin: Temporal Structures in Modern Society." *Social Research* 43 (1): 130–152.

Luhmann, N. 1992. *Describing the Future. Observations on Modernity*. trans. W. Whobrey. Stanford University Press.

Mahr, B. 2004. "Das Mögliche im Modell und Die Vermeidung der Fiktion." In *Science & Fiction: Über Gedankenexperimente in Wissenschaft, Philosophie und Literatur*, edited by T. Macho and A. Wunschel, 161–182. Frankfurt am Main: Fischer.

Mahr, B. 2011. "On the Epistemology of Models." In *Rethinking Epistemology*, edited by Günter Abel and James Conant, 301–352. Berlin; New York: de Gruyter.

Mahr, B. 2015. "Modelle und ihre Befragbarkeit: Grundlagen einer allgemeinen Modelltheorie." *Ewe* 26: 329–342.

Manifesto - The Breakthrough Institute, ed. 2015. *An Ecomodernist Manifesto*. www.ecomodernism.org.

Meadows, D. H., J. Randers, and W. W. Behrens III. 1972. *The Limits to Growth: A Report for the Club of Rome's Project on the Predicament of Mankind*. New York: Universe Books.

Merton, R. 1948. "The Self-Fulfilling Prophecy." *The Antioch Review* 8 (2): 193–210.

Nick, P., R. Fischer, D. Gradl, M. Gutmann, J. Kämper, T. Lamparter, and M. Riemann. 2019. *Modellorganismen*. Berlin: Springer Spektrum.

Nordmann, A. 2007. "If and Then: A Critique of Speculative NanoEthics." *NanoEthics* 1: 31–46.

Nordmann, A. 2014. "Responsible Innovation, the Art and Craft of Future Anticipation." *Journal of Responsible Innovation* 1 (1): 87–98.

Pielke, R. 2007. *The Honest Broker: Making Sense of Science in Policy and Politics*. Cambridge University Press.

Poznic, M., and R. Hillerbrand. 2021. "Scenarios as Tools of the Scientific Imagination: The Case of Climate Projections." *Perspectives on Science* 29 (1): 36–61. 10.1162/posc_a_00360.

Roßmann, M. 2020. "Vision as Make-Believe: How Narratives and Models Represent Sociotechnical Futures." *Journal of Responsible Innovation* 8 (1): 1–24. 10.1080/23299460.2020.1853395.

Roßmann, M., and C. Rösch. 2019. "Key-Narratives of microalgae Nutrition: Exploring Futures through a Public Delphi Survey in Germany [in Press]." *Science and Public Policy* 47 (1): 137–147. 10.1093/scipol/scz053.

Sand, M. 2019. "On 'not having a Future'." *Futures* 107: 98–106.

Schwemmer, O. 1976. *Theorie der rationalen Erklärung*. München: Beck.

Selin, C. 2008. "The Sociology of the Future: Tracing Stories of Technology and Time." *Sociology Compass* 2: 1878–1895.

Silvast, A., E. Laes, A. Simone, and G. Bombaerts. 2020. "What Do Energy modellers Know? an Ethnography of epistemic Values and Knowledge Models." *Energy Research & Social Science* 66. 10.1016/j.erss.2020.101495

Simon, H. A. 1957. *Models of Man: Social and Rational: Mathematical Essays on Rational Human Behaviour in a Social Setting*. New York: Wiley.

Stirling, A. 2008. "'Opening Up' and 'closing Down': Power, participation, and Pluralism in the Social Appraisal of Technology." *Science, Technology, & Human Values* 33 (2): 262–294.

Toon, A. 2014. *Models as Make-Believe. Imagination, Fiction and Scientific Representation*. London: Palgrave Macmillan.

van der Burg, S. 2014. "On the Hermeneutic Need for Future Anticipation." *Journal of Responsible Innovation* 1 (1): 99–102.

Walton, K. L. 1990. *Mimesis as Make-Believe. On the Foundations of the Representational Arts*. Cambridge, MA: Harvard Univ. Press.

11 Machine Hermeneutics

Alfred Nordmann

Hermeneutics is a practice of reading that constitutes the meaningful world of a text, be it a prophetic or biblical, even legal text, be it a literary work. The world we constitute through our hermeneutic practice is and isn't our world. It is our world in that we invest ourselves by making meaning and thus by constituting it, and it is not our world because it is the world of a text that resists appropriation. It cannot be integrated seamlessly into the experiences, ideas, expectations of our daily lives and cannot be absorbed entirely and does not dissolve into our ordinary ways of sense-making. The world of a literary work never becomes quite familiar but retains a sense of strangeness, unfamiliarity. In the meantime we inhabit a life-world which also is and isn't our world. It is ours because we live and act in it, it is familiar. It is not ours because we were cast into it and it is co-produced by countless human and non-human actors, with an uncertain future over which we have only very limited control even in our private lives. This meeting, blending, fusing of different, even antagonistic literary and life-worlds makes for what one might call the hermeneutic experience of the encounter with a text. Through this experience we change, we ourselves become someone else in our life-world and in the world of the text. We are so deeply implicated in this encounter that we cannot withdraw to the safe place of the observer or interpreter who casts out a net and retrieves some kind of model of what is said in a text and what it means. For the purposes merely of interpretation the reader is the measure of all things, recovering meaning on his or her own terms. In contrast, the reader is subject as well as object, agent as well as patient in the hermeneutic process – when I read a literary work, philosophical or legal text, the text happens to me just as much as I happen to the text.[1]

If hermeneutics encompasses all of that, what does this signify for "hermeneutic technology assessment" (Nordmann and Grunwald, this volume chapter 2)? On first sight, it can only refer to texts, that is, to the technological visions as they are articulated by advocates and critics of emerging technologies, or perhaps to the societal discourse at large. Treating technological discourse in analogy to a literary text, we discover

DOI: 10.4324/9781003322290-14

how meaning is made and what this tells us about ourselves. By choosing to read government policies, calls for proposals, ethics reports, TV documentaries, NGO position papers as if they were literally texts, we become implicated in a societal conversation revolving around anxiety and hope, prophetic confidence and doubt about the technologies of the day.[2]

It is not, however, the exclusive privilege of texts that they constitute a world for a reader to enter. Technical works are worlds in their own right, and we enter them as well. Of course, these include needleworks, artworks, musical and literary works as technical achievements, poetically brought forth by human ingenuity and labor. If works are worlds, this includes no less the world of the clockwork, the world of metal works or water works, that is, the world of the factory. There is now also, very prominently, the world constituted by electronic as well as social networks, that is, the worlds that correspond to socio-technical systems. To the extent that every technical process or device draws together material things as well as human developers and users, they can all be considered as worlds that result from the composition or putting together of numerous elements or parts. Do we enter these worlds as readers do a text? This is a somewhat speculative question that does not bear a literal answer. Readers or not, we seek orientation, hope to establish or maintain a sense of control, and contemplate the meaning of these works: Cosmological questions as to who we are in the world are raised by Dürer's *Melancolia*, Picasso's *Guernica,* or the *Wrapped Reichstag* by Christo and Jeanne-Claude, but no less so by a 16th-century astronomical clock that shows seconds and minutes, hours and days, months and years, and that exhibits the heavenly mechanics of the sun and the moon and the planets in their orbits. As practical an instrument as such a clock might have been, it is an object of contemplation as well.[3]

Hermeneutic technology assessment might be concerned, therefore, with the world of technical works – what they signify about the ways in which humans and things can live and work together. When archaeologists find a vase or bronze axe, they begin to reconstitute the ways of life and modes of production which might have resulted in this or that particular glaze. They come up with a world that is dissociated from our own world of daily lived experience – and it is important not to assimilate these separate worlds as they can inform and critique each other. In the temporal order, a prototype is the complement to the archaeological artefact. It is also not of our time, and though we can see it right before our eyes, it is emphatically not of our world. If archaeological artefacts are no longer part of our world or only as relics, prototypes are not yet part of our world or only as prospects. Similarly, the world of a prototype – of an artefact that supposedly heralds a future world – is not a mere extension or projection of our present world. Nor does it signify a latent world which only awaits to be realized. By claiming a way of being and doing things, these prototypes signify an alternative world which stands in an uneasy, as yet unresolved relation to our present world as we know it (see Dickel, this volume). How the present connects to the world of the archaeological artefacts is a question of hermeneutics, of telling a story which does not represent "the past" but constructs this pathway and connection and thus implicates ourselves – we change and become someone different in the telling of this story.[4] And how our present world connects to the world signified by the prototype is also a question of hermeneutics, of telling a story which does not represent the future but claims a prospect – not a possible future world but a model for reenactment, a world for the making. If the prototype is a model for enacting in the real world what so far is only a construct or concept, this invitation and prospect implicates us.[5] We change and become someone else by accepting or declining the invitation, though who or what we might become lies far beyond the design ambitions that produced the prototype.

Machine Imaginaries

The following pages provide a case study of hermeneutic technology assessment.[6] It exemplifies not just the process of entering the sociotechnical world of a work, and not just the process of seeking orientation or meaning in the unresolvable tension between the established and habituated world of daily experience and the structured, idealized sociotechnical world of a system or device. This case study deliberately seeks out prototypical machine concepts which do not insinuate a future that only awaits to be actualized, and which are not simply extrapolations from our present world. These concepts claim machines that are as yet unheard-of, beyond comprehension, perhaps impossible – not unlike the perpetual motion machine which might serve as patron saint for the following four. Like the *perpetuum mobile* these machines do not or did not exist literally or physically, and perhaps will never exist, but they already work, provide a model, a measure, an ideal for an imagined integration of people and things. Though they are conceived in technological terms they are inherently social.[7]

In the 1950s already, Günther Anders pointed out that our ways of valorizing technology express a sense of deficiency or vulnerability and, thus, *ex negativo*, a conception of a better life, of better humans in a better society (Anders 1956). Half a century later, Sheila Jasanoff foregrounds sociotechnical imaginaries: all the stories of technological determinism or enablement, including the visionary expectations of emergent technologies, are framed by sociotechnical imaginaries (Jasanoff 2015). These are imaginaries of how things can work together – of individuals working together in a social body, of component parts working together in a technological system.[8] Accordingly we encounter technical ideas of regularity and efficacy that prefigure conceptions of the economy or the state, and, inversely, machines as emblems of cooperation, of a governance of parts.

Sociotechnical imaginaries can valorize known technologies and can form an idealized image of a technology that is different and better than any known technology: We learn to view technology in the image of the social by way of conjuring up imaginaries of a technology that somehow ceases to be technology-as-we-know-it and becomes something else, unheard-of, better.

A horn of plenty or cornucopia, the lamp of Aladdin, the fairy tale's *Tischlein-deck-dich*, or nanotechnology's molecular assembler – these are wish-fulfillment machines that do the work of our dreams, that is, of dreams objectified in a public space, constituting a tractable order of things. Only somewhat less obvious than these, the soft machines of nanotechnology, their ancestors and distant cousins represent not the rationalization of the world but its secular re-enchantment. This strikingly nontechnical machinery is wonderful, indeed, and how things work in this enchanted world calls into question our familiar and mundane arrangements. The soft machines to be discussed do not constitute a technological paradigm that then infiltrates biology or social relations. Instead, they are first of all the collective dream of a technology that is unprecedentedly different – magically and wondrously so – from anything that was thought to be realizable so far. Soft machines are social machines insofar as they provide an escape from the limited technology currently in existence, an escape into the imaginary world of a technology that offers consolation and hope and is designed to establish a kind of peace throughout society. So soft and gentle are these machines that every hard, rigid, or constraining technology yields to them completely – including our as yet far too rough-hewn regulatory policy tools. So soft and gentle are these irreal devices that they are able to cushion the impacts of all conflicts of interest and social antagonisms and accommodate all "stakeholders" quite comfortably.

What will be pursued in the following, then, is not so much how a notion of the technical comes to be inscribed in different social domains but rather how a socially powerful notion of the technical emerges in the first place. Quite possibly, faith in technology is not an expression of confidence in the capacity of future technologies to solve at some time or another all our problems. As Astrid Schwarz has shown in her discussion of "green nanotechnology," the notion of green technology works as a hollow phrase precisely because it lacks or even defies precise technical meaning, giving it a highly effective symbolic function in public discourse by way of its ability to accommodate various societal visions. Whereas the technologies themselves are only in-the-making with uncertain success, they work now to quite effectively accommodate and coordinate interests and aspirations. And whereas the technologies-in-the-making involve us in zero-sum-games regarding social and environmental costs and benefits, they already provide a win-win situation for the many social actors who are willing to subscribe to its promise:

> The openness of discourse on the environment is used, then, to develop green nanotechnology as a space of possibilities compatible with societal wellbeing and sustainable in relation to nature. This is where the image of a kind of nanotechnology arises that is soft—or at least cautious—, and that reduces stark oppositions. It draws together what is thought to be irreconcilable, for example, natural history and physical reductionism, application-oriented basic research and scientific progress, the preservation of natural and cultural resources, limitation and transgression, scarcity and abundance.
>
> (Schwarz 2009, 117)[9]

This emergence of the technical in the space of the social invites hermeneutics, not only of the discourse around a concept, but beyond that also of the systems and devices that effect the transformations or reconfigurations of the natural and cultural, or the antagonism of people and things, of limitation and transgression, scarcity and abundance.

Lewis Mumford's Megamachines

Nontechnical social machines serve as imaginary ideals of both societal and technological development, they thus serve as sociotechnical imaginaries. Technology can be "nontechnical" because technology is not primarily an ingenious way of extracting as much as possible from the limited resources of a limited world. Instead, nontechnical dreams of technology envision that the world could turn out to be limitless, after all, and that technology can alter our conception of what is technically possible. While Ernst Cassirer draws a strict dividing line between this magical image of technology and the realities of engineering in the context of nature and society (Cassirer 1930, 59–60), certain machines overcome these differences as they transform our familiar world.

According to Lewis Mumford, fantasies of technical control assume that technology can advance "without encountering either technical delays or sobering human inhibitions," thereby turning "our dominant technology itself into the equivalent of scientific fiction." About technology as social and scientific fiction, Lewis Mumford goes on to say, "The one part of the human personality that so far eludes rational control is that which produces these fantasies" (Mumford 1970, 223 & 290). Humans, he maintained, technologically transformed the conditions of their lives by calling upon the machine, and now they seek to escape the paradigm of control and imagine a new configuration of humanity and technology, one that produces a new way of being human. To understand what Mumford means by this, it is worth taking a closer look at the argument he presents in his book *The Myth of the Machine* (1967; 1970).[10]

According to Mumford, "[W]e cannot understand the role that technics has played in human development without a deeper insight into the historic nature of humans" (1967, 4). Humans express through technology their latent potentialities "to fulfill more adequately [their] superorganic demands and aspirations" (ibid., 8). The needs and desires that are superorganic here are those that are associated with human freedom – a familiar idea, philosophically speaking. Accordingly, the "dominant human trait" is the capacity "for conscious, purposeful self-identification, self-transformation, and ultimately for self-understanding" (ibid., 10). The self-understanding of humans as historically changeable beings occurs by the very means of technology itself, namely, in the encounter with a world that has been deliberately and consciously transformed by human hands. And what we become aware of when we see ourselves in our technology is the form of societal organization that precedes the reification and the concrete manifestation of technology. Thanks to a division of labor that is organized in a specific way, that is, with the help of technology, humans create for one another a world of things in which they recognize themselves – and in which they may be able to transform themselves.

In this deliberately created form of organization, the "megamachine" makes its appearance. To define the megamachine, Mumford quotes from the standard engineering text of the late 19th century, *Kinematics of Machinery* by Franz Reuleaux (1875), which proved to be extraordinarily influential also among philosophers. According to Reuleaux and Mumford, a machine is a combination of resistant parts, which tames or domesticates the forces that enter in and are channeled through it so as to perform a well-defined choreographed movement. It thus operates under human control to utilize energy and to perform work (Mumford 1967, 191; cf. Reuleaux 1875, 38). Mumford claims that the history of machines which meet this definition goes back at least five thousand years. At that time, the machine was not yet visible as a device but existed solely as a form of societal organization that laid the foundations for the visible machines of the 19th and 20th centuries: "[T]he mechanical agents had first to be 'socialized' before the machine itself could be fully mechanized" (Mumford 1967, 194).

It was this invisible – in a material sense as yet nontechnical – working machine that enabled the pyramids to be built five thousand years ago. Though "made of human bone, nerve, and muscle," the components of the machine were, by definition, "reduced to their bare mechanical elements and rigidly standardized for the performance of their limited tasks. [...] The secret of mechanical control was to have a single mind with a well-defined aim at the head of the organization, and a method of passing messages through a series of intermediate functionaries until they reached the smallest unit" (Mumford 1967, 191–192). For Mumford, this social machine is not the societal manifestation of a technical ideal; rather, it exists prior to the construction of the mechanized machine as an engineered device. He envisages the free intellectual invention of a form of organization that already knows how to utilize mechanical forces before it becomes materialized in a fully mechanical labor-saving machine.[11] In the case of the pyramids and a strict, perhaps totalitarian organization of labor, mechanical forces were still socialized rather than mechanized; as such, they were part of a societal mechanism rather than a technical one. The model for the interplay between cogwheels, screws, winches, and levers comes from this interaction between humans at work – to borrow from Foucault, we see in the social and technical organization of labor a working order of things.[12] For Mumford, then, when people recognize themselves in the machine, they are not seeing in it their biological or individual nature but the way their society or the megamachine is organized, and, in this organization, they are themselves only part of the system. And when these inhabitants of a machine culture adopt the perspective of liberty, they are able to turn against this megamachine by means of conscious self-identification and self-transformation.

Hermeneutically speaking, Mumford thus takes the machine as a work and world, enters into it as a socially organized world that signifies the reign of the mechanical, requiring a kind of resistance in the name of freedom, challenging us to differentiate between mechanized and socialized forces.

Otto Mayr's Centrifugal Governor

Within the philosophy of technology, Mumford's is perhaps the most significant and comprehensive version of the theory that technological paradigms are constituted by society long before the corresponding mechanical device is developed. Moving on from this, we now look at a more narrowly defined social machine, introduced here to act as a foil for "soft machines."

As curator of the National Museum of History and Technology in Washington, D.C., historian of technology Otto Mayr wanted to write a history of the centrifugal governor, a mechanism for automatic regulation that plays a crucial role in the safety of technical systems. Steam engines produce pressure in a boiler that then drives a transmission system. High pressure in the boiler carries with it the danger of too much pressure building up, leading to an explosion. This is where the centrifugal governor comes into play. It consists of a few metal "flyweight" balls arranged as a kind of carousel that is propelled by steam pressure. The higher the pressure in the boiler, the faster rotates the carousel, and the more quickly it rotates, the stronger the centrifugal forces acting on the spinning flyweights: these rise higher and higher as the rotational speed increases until they eventually reach a horizontal position. Once the steam pressure in the boiler has reached a critical point, and with it the speed of the rotating flyweights so that the latter have reached a nearly horizontal position, the governor opens up a valve and steam is able to escape from the boiler. This causes the steam pressure to decrease, the carousel to rotate more slowly, and the flyweights to sink back down. The valve now closes again so that only the right amount of steam escapes. At this point, the whole process can start over again, with the centrifugal governor acting as an automatic system of regulation without which the steam engine and the Industrial Revolution would not have been possible. This is a system of regulation, moreover, that works with a so-called feedback loop, as its behavior depends on the system of which it is a part, that is, on the system that it is designed to observe and on which it acts in return.

When he embarked on writing the history of this system, Mayr made an interesting observation: centrifugal governors had been described in a famous list of machines dating from the year 60 that had been reprinted in the 16th century and had become the blueprint for various mechanical showpieces, not least among them a host of clockwork-driven devices. However, well into the 18th century, there are no feedback mechanisms to be found in continental Europe but only in Britain where the technology had been "cultivated and appreciated." Why was this the case? Mayr's search for an answer led him to conclude that it was not possible to explain the rejection of these systems in continental Europe and their development in Britain by reference to any developmental trajectory inherent in to the technology itself:

> I had noticed that, in eighteenth century Britain, the principle of the feedback loop had come into use not only in practical technology but also in abstract arguments, notably in Adam Smith's economic theory. Assuming that this was not coincidental, I tried to establish a connection. I tried to show that the use of the concept in abstract argument had been inspired by practical technology. The attempt failed; in the end, I became convinced that the connection was not direct but that each phenomenon independently was the result of some unknown earlier cause. This suggested that I take a closer look at the first question: Why was feedback rejected on the Continent?
>
> The question could not be answered in terms of developments in practical theory alone. [...] Perhaps the answer was to be sought on another level. Apparently, certain kinds of mechanical inventions were immensely popular on the Continent, while others, including feedback devices, were not: What made the difference?
>
> (Mayr 1986, xvi)

From here, Mayr's book goes on to claim that the technical development of clockwork mechanisms and of feedback systems was due to interactions "between the political, social, economic, or religious ideas dominant in a given society and contemporary preferences and designs of technological hardware" (Mayr 1986, xv). Viewed from this perspective, Adam Smith's economic theory precedes the dissemination of the centrifugal governor in mechanical devices not just in chronological terms: the invisible hand that balances supply and demand finds its counterpart in the automatic action of the mechanism that maintains balance in the steam engine. Thus, Mayr refers to feedback mechanisms as "liberal systems," whereas the clockworks of baroque continental Europe reflect an immutable authoritarian system that is hierarchically organized. To imagine a liberal state is to promote self-regulation; accordingly, liberal economic theories appeared prior to technical refinements of the governor – which, in turn, enabled the governor to become a kind of technical metaphor for the modern state.

> The steam engine governor probably did more than any other agent to publicize the concept of self-regulation among engineers and the general population. The Watt steam engine was greeted as a machine of revolutionary importance and as the herald of a new age. No one would miss an opportunity to see this wonder in operation, and few who saw it would have failed to inquire about the purpose of those rapidly rotating centrifugal weights that were mounted conspicuously over the machine. To explain the concept of self-regulation, from that time on, one only had to point out the steam engine governor. When Norbert Wiener in 1947 christened his new science of cybernetics, he was expressly paying tribute to what he considered the earliest cybernetic device; the word *governor* is derived via the Latin *gubernator*, from the Greek [for] steersman.
>
> (Mayr 1986, 194–195)

A modern-day observer looking at the steam engine designed to operate the water fountain in the royal gardens of Potsdam, for example, will see – despite its rather weak performance by today's standards – a technological miracle housed in an exotic temple, high up on which the bronze governor sits enthroned, a polished and gleaming wise steersman or navigator. And by no means accidentally, atop the regulator as an emblem of enlightened governance sits the crowned Prussian eagle. Likewise, to the present day, the art of government in a liberal state relies on a regulatory system. The material quality of products, the concentration of toxins in exhaust fumes, and the ingredients in food items are all subject to regulation, in many cases enshrined in law. There are established routines, if not automatic procedures, for observation, and corrective measures that are taken once a defined threshold is transgressed. This system is motivated by the notion of balancing the innovative strength of the market against consumers' safety needs. If anything bad were to happen, the measure kicks in and balance is quickly restored.[13]

Again, hermeneutically speaking, it is not that technical feedback and control gives rise to a system or notion of governance, but the technical device is deployed as a liberal technology, as a technology of self-maintenance, of balancing interests and forces, as an image of social relations.

Jessica Riskin's Wetware

Mumford's working machine and Mayr's centrifugal governor have served as examples to show that societies imagine technologies even before the latter actually exist in the form of specific devices or technically realized machines. These machines are at first completely social, materializing only at a later time; they are not inspired by technology. And it can be said that the regulatory control system with its automatic feedback mechanism and its balancing function was a paradigm for the development of society and technology alike.

No matter how complex the interplay between ideas of a social and a technical order or people and things, it has given rise to a highly tangible technological world. Not only are there now very real factory floors on which work machines are organized into production lines, but also boiler safety technologies with their theoretical underpinnings and engineering practices. However, a more far-reaching claim was put forward at the beginning of this chapter, namely, that societies also find orientation in a quite nontechnical idea of technology, that is, in regard to technologies that do not exist, perhaps cannot be materially realized. This will be shown in reference to *soft machines*, but this discussion, too, can be prepared by way of historical precedent. Jessica Riskin describes a short-lived fanciful technology that briefly appeared during the second half of the 18th century and that foreshadows a similarly precarious development in the current age of soft machines.

Only recently in Brussels, window shoppers might have encountered a rather large cat stretched out luxuriantly on its back amid piles of Belgian chocolates. Pretty soon they would have realized that the cat is a mechanical toy performing certain cat-like movements. It strokes its whiskers with its paw as a contented swell rises and falls across its thick underbelly. Jessica Riskin wrote a history of such devices, and, in her words, this toy functions analogically. Inside is a kind of clockwork that functions like a mechanical model. It is set up to represent and produce certain visible signs or movements of a cat. The very fact that we recognize it as a toy and understand the representational character of the mechanical model makes it clear that this is an instance of a piece of hardware that technically mediates the gap between a real biological cat and the mechanics of its appearance. This hardware has little in common with the internal living being of an actual cat, except that it is able to reproduce certain visible signs, albeit using quite different means than those of a real cat. According to Riskin, there were legions of such automata in existence in the 17th and especially the 19th centuries.

However, between 1730 and 1790, the focus was on other constructions, which Riskin, in contrast to analogical *hardware*, describes as *wetware*, thereby using a very contemporary term from artificial intelligence research and biocybernetics.

Riskin illustrates the difference by describing various writing automata, that is, life-like dolls that move a feather quill across a sheet of paper. In the 17th and 19th centuries, such automata made it seem as if they were writing, moving the quill in characteristic manner through the air above the paper. In the 18th century, however, the aim was to make the apparatus actually write for real. And indeed a writing figure originating from the workshop of Jaquet-Droz was able to write up to 40 programmable words. The inside of this figure looked different as well:

> The Jaquet-Droz automata do not just carry out the processes of writing, drawing, and playing music, they are also anatomical and physiological simulations. Their skeletal structures were likely designed with the help of the village surgeon. Both the Lady-musician and the Draughtsman also breathe. The Draughtsman periodically blows the charcoal dust from his paper and surveys his work, and the Lady-musician sighs in time to the music. Her breathing was what spectators most often commented upon. It made her seem not only alive, but emotional. She appeared moved by the music she played.
>
> (Riskin 2003, 102)

So these are not simply mechanisms that model and represent certain behaviors analogically; they are simulations of living behavior, extending as far as the birth machines of Madame du Coudray which were constructed from pelvic bone, fabric, leather, and moist sponges and served quite literally as wetware so that midwives could rehearse something akin to the process of birth. That these simulations were crude and clumsy does not defeat their purpose – these 18th-century machines sought to learn from simulations rather than from analogical models.[14] Today's biocyberneticians, intelligence and behavioral researchers describe the organic connections in the human brain as "wetware" in contrast to the hardware of a computer. To be sure, one can use the latter to imitate intelligent behaviors or thought processes in a technological medium, yet the fact that an analogy must first be established between organic and electrical processes signifies a gap between model and world, construction and reality. In the simulation of living beings, no such gap should even appear. Thus, a specific demand for knowledge calls for an intimate proximity to things. And this epistemological requirement entails an expectation that the machinery of simulation cannot actually fulfill.

The automata of the 18th century exhibit clearly how far they fall short of this unrealistic expectation. They are unable to simulate living processes in such a way as to generate theoretical insights or practical skills regarding the physiology of humans and animals in an unmediated way, directly from beholding the machine. The question arises, though, whether today's simulations have really come much farther than that or whether they still gesture clumsily at an unachievable ideal of technology – an ideal device that has been assigned a definite role in the research process and that already functions as a public object of fascination. Riskin comments on this:

> "Wetware" … is the expression of a particular moment, the turn of the twentieth to the twenty-first century. The neologism voices one of the organizing ambivalences of the current moment: we believe that the processes of life and consciousness are essentially mechanistic and can therefore be simulated, and yet we are equally firmly persuaded that the essences of life and consciousness will ultimately be beyond the reach of mechanical reproduction.
>
> (Riskin 2003, 97)

As described by Riskin, this ambivalence is an expression of a profound confusion and perplexity that would leave unjustified any expectation of a kind of technology that somehow transcends its mechanistic origins. The simulations that had made an appearance in the 18th century disappeared again in the 19th because the idea of such a technology could not be developed coherently; and they are now returning because, in the transition from the 20th to the 21st century, we find ourselves once again confused and perplexed. This, of course, raises the further question of why these two historical moments, in particular, should be characterized by such confusion and perplexity.

Riskin picks up on this question: "Why were the seventeenth and nineteenth centuries periods of analogy, and the late eighteenth and late twentieth centuries periods of simulation?" (2003, 118). By way of an answer she suggests that analogies are employed when people consider what machines are and find that animals are a lot like them and can be described in the language of machinery. In contrast, simulations respond to perplexity when people are not entirely sure what animals are, and equally unsure what machines can do, and seek to find out about both by trying to build an animalmachine. With their animalmachine, then, the researchers of the 18th and 21st centuries are imagining a technology without knowing whether it can even exist, that is, whether its very idea may turn out to be incoherent. As such, these machines are an especially vivid example of the fact that technological ideals can be, as it were, nontechnical, indeed almost antitechnical:

> The beginnings of the Industrial Revolution and the beginnings of the Information Revolution were both periods of extreme fluidity in people's understandings of what machines were—and indeed in the

> nature of machines. Once the industrial period was fully under way, manmade machinery and its relations to living creatures stabilized, replacing the fluidity required by a simulation with two terms that were, for the moment, fixed: "life" and "mechanism." Only when the Information Revolution introduced a new kind of machinery did this fixity give way to a new fluidity, and the possibility of using machinery to simulate life again seemed intriguing. In other words, the modern makers of automata that see, hear, and feel in fact have a great deal in common with the eighteenth-century makers of automata that breathed, spoke, and defecated. They too use machines to simulate life precisely because their conception of machines is no better established than their understanding of life.
>
> (Riskin 2003, 118–119)

With regard to the claim being developed here, then, one might say that societies hope for orientation from merely imagined technology because they understand neither themselves nor technology and because they are troubled, fascinated, and perplexed all at once by the limits and possibilities of technology and by the future promise of an unprecedented new and different technology. Such perplexity is provoked by the life-simulating wetware of the pre- and postindustrial age, also by AI-driven humanoid robots or by the "soft" social machines that are imagined in our own times. As we imaginatively enter the world of an imaginary kind of machinery, we are looking for a transformative moment, and it matters not that the presumed work of technology might be a work of fiction. The science/fiction distinction itself becomes erased in the discourse of emerging technologies, and reappears only when we notice that the desired transformation is not really that of technical practice and the material order of things but primarily one of the social order and the alignment of people and things.

Richard Jones's Soft Machines

The succession of historical examples has now arrived in the very recent past and affords a view of nanotechnology as a prime example of so-called emerging technologies. Against our own better judgment, emerging technologies feed the hope for solutions to nearly all pressing problems – nanotechnology, synthetic biology, geoengineering, AI, genome editing promise cures for ageing and disease, sustainability and resource depletion, pollution and climate change. Nanotechnology is an especially interesting case because at its beginnings stood the idea of an unprecedented machinery, and, since its beginnings, the possible realization or likely impossibility of these machines were at issue. Nanotechnology's initial claim to fame rested on things called "molecular assemblers" or "nanobots," which may either exist sooner or later or else are completely impossible.[15]

The highly differentiated nanomachine discourse cropped up in many different areas of society, including cultural and social science as well as philosophical analysis, but cannot be discussed here (see Drexler 2003a, 2003b; Lösch 2007; Myers 2008; Nerlich 2005; Smalley 2003a, 2003b). While Bernadette Bensaude-Vincent and Xavier Guichet (2007) draw an analytic distinction between Cartesian, complex and concrete machines, the following takes as its starting point the popular book *Soft Machines* by physicist Richard Jones (2004) with its distinction between hard and soft machines. Interestingly, Jones's soft machines are not based on any prior technological model. He introduces them not because he sees prototypes, proofs of concept, and a steadily advancing technical agenda. Instead, they fill a conceptual void. If one wants to make sense of the visionary notion of "molecular machines," soft machines are the only conceivable candidate for such machines – scaled-down "hard machines" won't do the job. Since molecular machines are not compatible with familiar technical paradigms, a new, quite different kind of technology – wet and soft and modeled on biological processes – needs to be imagined on which, in principle, these machines could be based. Jones presents all this with great intelligence and care – he knows exactly what he is doing. As a physicist, he does not assume any obligation to prove the technical feasibility of the soft machines he conceptualizes. Indeed, even as he explores the idea of what a molecular machine would have to be from the point of view of physics and biology, his book is ultimately one on the relations between nanotechnology and society.

The question of nanomachinery is salient, according to Jones and numerous other researchers, because it is on this question that societies might refuse to accept nanotechnology's beneficial nature. An understanding of "soft machines," Jones argues, provides a necessary corrective to dangerously simple-minded ideas about the ways in which nanotechnology will disrupt social relations. Those who speak of "molecular machines," "nanobots," "assemblers," and the like are aligning themselves with the visionary rhetoric that was initially introduced by Eric Drexler. According to this rhetoric, nanotechnology establishes control over individual atoms and molecules that is so precise as to enable the construction of tiny, scaled-down, otherwise ordinary machines that will perform work, for example, by fabricating artefacts, repairing human cells, or clearing arteries. For Jones, this notion is dangerously misleading for various reasons. Given that it is not scientifically credible (Jones 2004; Smalley 2003a, 2003b), it gave nanotechnology a bad reputation and undermined all its promises. And even though it is not scientifically credible, it might, nonetheless, trigger irrational fears of uncontrollable nanobots.

That this notion lacks credibility, however, opens up the possibility of a thought experiment – one that leads away from the idea of scaled-down conventional machinery at the molecular level and that imagines instead an unconventional kind of machine that, although having fine scientific credentials, is beyond technical reach. Jones asks what kinds of machines could perform work at the nanoscale. These would be molecules among molecules in the bloodstream, for instance. A machine of this sort would have to know how to utilize the particular conditions of a world in which gravitational forces are practically absent and where Brownian motion dominates. A machine designed according to the principles that rule the macroscopic world would stand no chance at all against Brownian motion. Instead, what would be needed is a biologically soft machine, for example, a viral or sperm-like machine that is adapted to its environment and that allows itself to be driven by its environment such that it is the environmental conditions which sustain its proper functioning.[16] In this way – and to counter the scientifically incredible nanobot – Jones brings into play a scientifically respectable, albeit merely hypothetical, machine that appears to take its cue from political philosopher Ernst Bloch. Decades ago Bloch dreamt of technology that was allied with nature to produce an "unprecedented insertion" and the "building of human work into the work of nature" (Bloch 1973, 817). In what sense "soft machines" can even be considered "machines" in the first place is not at all clear; neither is it clear whether they can ever be realized in technical terms. Nonetheless, Jones's soft machines promise something new that eludes the opposition of natural and social relations. Instead, these machines are integrated into the physical and social dynamics of human affairs and natural processes – they make for a sensitive, considerate, conducive, friendly technology.

It may come as no surprise, then, that physicist and nano-researcher Richard Jones not only wrote this book but subsequently became one of the most prominent and effective nanotechnology communicators in Britain. Influenced by critical scholarship in the field of science and technology studies, Britain had rejected the simple model according to which the communication of scientific ideas serves primarily to promote their acceptance. Here again, what was intended instead is an "unprecedented insertion" and a notion of governance as building or designing human work into the work of science and technology (Kearnes et al. 2006). The State should no longer be conceived as a hard machine to whose regulatory agencies citizens address their concerns merely by submitting petitions and inquiries, or as a machine with a rigid funding mechanism that is tied to immutable criteria of scientific quality and economic profitability. Rather, processes of technological development should prove to be gentle and soft, open to collective design, accommodating and trustworthy. In this sense, Richard Jones filled his officially appointed role of ombudsman for nanotechnology in an exemplary manner. These were exemplary not only on account of his analyses, which are readily accessible, carefully differentiated, clearly understandable, and often funny, and not only because he was crucially involved in public consultations aimed at establishing research priorities. Jones also filled his role in exemplary fashion in that he engaged in dialogue with scholars in the social sciences and the humanities (for example, Jones 2011). As a representative or model citizen of socially constituted research on emerging technologies – molded after the notion of "mode 2 knowledge production" with its new social contract (e.g. Nowotny et al. 2004) – Jones constructed two kinds of soft machinery and two technologies so gentle as to be unreal. His biologically inspired nanoscale machine and the participatory governance machine mutually inform one another, since both share in the social imaginary of a never-to-be-realized but altogether gentle and accommodating working order of things.

Jones's soft or biological machines will not be considered as machines by some, for the simple reason that the blurring of the organic and the mechanical constitutes a category mistake – especially if to the organic is attributed something akin to emergence or spontaneity and, thus, something that eludes predictive control. Jean-Pierre Dupuy, in particular, has emphasized that nanotechnological research programs appear to be dedicated to the paradoxical task of controlling the uncontrollable. This is exemplified by the appropriately muddled words of an anonymous nano researcher:

> The problem is the illusion of control—what we want to do is reverse engineer. We harness self-assembly in a non-linear way to get what we want. To do this at the nanoscale will be a big breakthrough because we can then start to control things, put them in compartments and let them evolve. We don't need the illusion of control. We let the system select what it needs according to its local environment. We can't be an engineer at that level if we want to use bottom up. Nature takes this approach and it works very well.
>
> (Quoted in Macnaghten and Kearnes 2007, 17)

The paradoxical brief of controlling without control includes the construction of "soft machines" and appears here under the heading of "bottom-up engineering" which refers to the use of self-organizing processes such that technological systems can grow and need not be constructed. Self-organization, contends Dupuy and herein echoes the anonymous researcher, is fundamentally contrary to the art of engineering, and he notes that a dynamic that is sometimes labeled as nonlinear dynamics and sometimes as chaos theory contradicts the idea of reliably functioning technological systems. Engineers, he says, surely have no intention of allowing their constructions to surprise them and surely will not build technical systems so complex that they feature catastrophic tipping points (Dupuy 2007).

Soft Governance in a Plastic World

The opposite of realism is not utopian but wishful thinking (for example, Geuss 2010). Accordingly, the genie of the lamp or the perpetual motion machine is no utopia either but rather a wish-fulfilling machine – not much different from the steam engine's regulator whose art of governance transforms individuals' interests into the common good, and also not much different from the intelligence of an organized division of labor that allocates each individual a place in the working order of things. One could go on from here to produce a critique of the discourse on machine enhancement and attendant concepts of a nontechnical technology that reflect wishful thinking but have no technological precedents. This would be one kind of "hermeneutic TA" – one that seeks to understand and critically assess visions of the future as they are produced in and indicative of the present. It would expose a confused state of debate along with a kind of wishful thinking that allows for inferences about what, presumably, is considered to be so deficient in the present as to require such desperate, fanciful measures (Nordmann 2010a).

There is another route to be taken for "hermeneutic TA" – one that engages with the envisioned working order of the imagined machinery. If ours is an age in which we do not know what humans are and what only we can do, and at the same time do not know what machines are and what they might be capable of, and if we are therefore unsure about the limits of technology, there is no easy way of dismissing wishful thinking and the conception of unprecedented machines. By taking them as sociotechnical imaginaries of how things might work together, they can neither be valorized for their normative expectation of a better world nor be dismissed as illusory and misleading – the conception of sociotechnical imaginaries cuts across the distinction between utopian thought and wishful thinking.[17] In sociotechnical imaginaries, historical experience appears in the horizon of hope and destiny, of what might be possible and what should be necessary. As such, these imaginaries are saturated with experience and "false consciousness," but, in irreal conceptions of unprecedented machines, they also create images of peaceful accommodation, the end of politics, or the infinite plasticity of the natural and social world. These images need to be recognized rather than debunked, rendered amenable for public deliberation. Irony and caricature can help, if only to loosen the hold on the imagination of human enhancement, for example. When the idea of perfecting human nature is discussed alongside that of enhancing machine nature and of enhancing material nature (Nordmann 2010b), these various ideas will gain mutual support by referring to a shared historical reality but, by the same token, will also appear more precarious – symptoms perhaps of normative uncertainty and cultural instability.

It would be a major surprise if it actually proved possible to build transparent wish-fulfillment machines that offered no resistance, that exerted control without exerting control, and that seamlessly expressed the order of nature and a social world.[18] At the same time, it is no surprise that the notion of controlling the uncontrollable inhabits people's imaginations and informs the machinery of social life. The very idea of technical control over historically contingent, dynamic processes offers the promise that politics need no longer be an art but can reliably manufacture social harmony and consensus. Indeed, this is what an imagined technology can facilitate and enable – the depoliticization of politics. In the nonlocal public sphere of soft law, soft regulation, and soft governance, there is no hard machinery of thresholds and regulatory interventions. Its social machinery seeks to exploit the effortless compulsion from self-organizing, almost organic processes of growth.

If one wanted to cite specific reasons for the emergence of such a governance conception of politics and technology, one could do worse than consider Jessica Riskin's diagnosis of our time. She has postulated a connection between the fascination with technical simulations of organic processes and people's sense of perplexity regarding the possibilities and limitations of technology. This perplexity can be interpreted as a response to the general notion that there are no limits to the world's plasticity or malleability – a notion for which nanotechnology and synthetic biology are only the most recent examples. Indeed, this plastic world appears to be dematerialized to such an extent that it has run out of ways to resist technological intervention (Bensaude-Vincent 2004). Thus, the non-physical, soft social machine embodies a broadened conception of design that draws together the work of developers and users, technicians, engineers, scientists, consumers, and citizens – if everyone is invited to contribute to the shaping of an infinitely plastic world, conflict will evaporate and environmental problems, for example, "will take care of themselves" (Nordmann and Schwarz 2010). Mumford, in particular, elaborated on the historical significance of a technology that initially existed only in the imagination. This is what we are seeing also today: The soft governance model of a collective social experiment with new technologies suggests the emergence of a social order in which producers and developers voluntarily agree to be accountable, in which consumers willingly act as guinea pigs, in which analytic expertise is spread among all participating citizens, in which monitoring by state agencies is replaced by permanent vigilance distributed over an indefinite number of actors. Subsequently, conflicts of interest become backgrounded in the discourse of responsibility, while "ethics" serves as an international *lingua franca*. All this, at any rate, is part of the dream of soft machines as emblems of environmentally benign, socially integrative technology.

Technical works – whether materially realized or an imagined working order of things – are worlds that are subject to hermeneutics, as such not unlike literary works, biblical or legal texts. We assess these works by trying them on as worlds – there is usually a fit like that between lock and key, problem and solution, but there are many more and less conspicuous misfits as well like that between machine intelligence and human understanding or that between new kinds of agency and old habits of the mind or habits of the law. There is no single criterion here and only first steps towards an accommodation with resistances on both sides as one enters the world of the socio-technical work. Trying on and thereby assessing a world, this is the subject of so-called scenario exercises. Depending on how such exercises are conducted, these well-established, old-fashioned tools provide at least a rudimentary form of hermeneutic TA. Importantly, all this has nothing to do with understanding our technological future. To be sure, visioneering and prototyping often posit an imaginary working order of people and things as a possible future – this or that scenario or imaginary socio-technical work is to be privileged as a special kind of world, namely a scenario or world of the future. But hermeneutic TA cares not how and why and by what right such privilege might be accorded. We do need to know. Whether we read historical novels, science fiction, or fantastic literature, the question is how we enter these worlds and how we learn to reflect our given everyday world through this encounter – whether we want to dismiss or pursue some brave new world. This also holds, so the argument went, for technical works, materially realized or imagined as a possible working order of people and things.

Notes

1 This opening takes a particular stance in respect to hermeneutics and hermeneutic TA. It aligns itself with Schleiermacher, Gadamer, Ricoeur, and others. The quest for "understanding" can be satisfied by explanations, background information, interpretation, and other sense-making activities. Hermeneutic understanding, however, requires more, namely an activity and ability to reproduce, re-enact or perform, to re-constitute, or participate in the world of a text.

2 Considered as such, "hermeneutic TA" is an extension of "vision assessment" (Grin and Grunwald 2000, Sand this volume) as one element in the larger toolbox of technology assessment.

3 This is echoed in a somewhat cryptic remark dated October 8, 1916 in just those notebooks of Ludwig Wittgenstein in which he drafted his *Tractatus Logico-Philosophicus*: "Habe ich aber den Ofen kontempliert, so war *er* meine Welt, und alles andere dagegen blaß. [So when I was contemplating the stove, *it* had become my world, and everything pale in comparison.]"

4 The construction of such a pathway can be referred to as "visioneering": "What visioneering aims for is to exhibit a compelling causal link between a state A (technological work-in-progress) and a state B (a future so desirable as to mandate its realization) such that A will actually lead to B while B necessitates A." (Nordmann 2013, 89; cf. McCray 2013; Sand this volume).

5 Compare Arie Rip "Technology as Prospective Ontology" (Rip 2009) and a more elaborate account of prospective modelling by Alfred Nordmann is in preparation for the volume "Modeling the Possible," edited by Tarja Knuuttila.

6 This is the third paper to draw on this case study. Earlier versions answered to different questions, see Nordmann 2014, Nordmann 2016.

7 For the perpetual machine or, rather, its antithesis – thermodynamics and the conservation of force – a hermeneutic analysis *par excellence* was provided by Rabinbach (1992).
8 Jasanoff (2015) quotes Charles Taylor on social imaginaries: "I am thinking … of the ways people imagine their social existence, how they fit together with others, how things go on between them and their fellows" (Taylor 2004, 23).
9 In June 2011 a series of vignettes published in the weekly paper *Die Zeit* highlighted the way in which an essentially non-technical concept of technology can serve as mere wish fulfillment to legitimize society's hopes of a last-minute rescue from, among other things, the effects of climate change. After characterizing today's techno-optimism as a kind of desperate hope, the authors present a range of fantastical technologies of the future under the heading "*We can be optimists after all: miracles happen all the time*" (Fischermann/von Randow 2011).
10 The remainder of this section is based on the discussion in Nordmann 2008, 53–56.
11 The temporal priority of a social idea of the organization of people and things in some technical setting should not be confused with a technological Platonism as it was advocated by Dessauer (1927). These social ideas are anything but "mere" ideas, they are efficacious and perform work, if only the work of the prototype with its "this is how" and simultaneously its "not yet."
12 To be sure, this is to borrow from the English title of Foucault's *Les Mots et les* Choses (1966). It also resonates with Joe Pitt's definition of technology as "humanity at work" (Pitt, 2011, 70).
13 The governor or cybernetic controller of the steam engine is not only a cipher of the "first industrial revolution" and not only the epitome of the technocratic "control and regulation" paradigm that was criticized by Heidegger and Marcuse. It continues to be influential in conceptions of "governance beyond the state" where it refers to self-regulatory mechanisms for the reconciliation of interests or the balancing of goods. Lösch et al. 2009 ask whether perhaps a social machine other than the centrifugal governor might be required as a political model – one that is more appropriate to current modes of technoscientific and industrial production. The question is all the more urgent in light of the fact that the development of soft machines in nanotechnology, synthetic biology, information and communications technologies eludes classical regulatory procedures. The fourth case study will show that it is to devise a new work as a new world when we imagine a different working order of people and things and new governance concepts along with it.
14 Riskin's distinction cannot be understood from the vantage point of current debates about digital and analog technologies: her "simulations" are analogical devices as well. Her contrast between simulation and analogy is more like the distinction between icon and index, where the iconic is based on the immediate participation of the sign in the signified (simulation) and an index corresponds to the two-place relation between the representation and that which is represented (analogy).
15 The fact that the allure of nanotechnology depended entirely on this dream was confirmed by our sobered awakening. As soon as the promise of these machines vanished from the generally accepted list of nanotechnological promises, nanotechnology lost its magic spell or fascination. It is now an R&D trajectory like any other and is for the most part normalized – no longer a matter of molecular machines and atomic precision but mostly of nanomaterials. Collective dreams of novel machinery are nowadays projected upon synthetic biology and AI.

16 To be sure, there are such machines. If we choose to consider them "machines," Jones's *flagellae* are among them as are Gilbert Simondon's concrete machines (Simondon 1958). Jones's argument concerns only the conceit that scaled-down *Lego*-like precision control could produce such biological machines at the nanoscale.

17 According to Jasanoff, "Sociotechnical imaginaries occupy the theoretically uninscribed space between the idealistic collective imaginations identified by social and political theorists and the politically neutered, hybrid networks or assemblages with which STS scholars tend to describe reality." These are visions, then, not of a future that departs from the present but of an idea of the future that reinforces the present. As such, "[u]nlike mere ideas and fashions, sociotechnical imaginaries are collective, durable, capable of being performed" (Jasanoff 2015). In the following, it will be shown in which sense even Jones's soft machines are collective and durable imagined technologies and that they are capable of being performed.

18 Eric Drexler's "universal assemblers" were quite explicitly to be wish-fulfilling machines and, as such, were able to influence the early years of nanotechnology research in a decisive way. Also, the dream of being able to translate mere thinking into a functioning machine makes for the fascination of so-called mind-machine interfaces (even if these will always be brain-machine interfaces).

References

Anders, G. 1956. *Die Antiquiertheit Des Menschen*. München: Beck.

Bensaude-Vincent, B. 2004. *Se libérer De La matière? Fantasmes autour Des nouvelles Technologies*. Paris: Institut national de la recherche agronomique.

Bensaude-Vincent, B., and X. Guichet. 2007. "Nanomachine: One Word for Three Different paradigms." *Technè 11* (1): 71–89.

Bloch, E. 1973. "*Das Prinzip Hoffnung*. Frankfurt: Suhrkamp." In English: *Principle of Hope*. (3 vols.) (N. Plaice, S. Plaice and P. Knight, Trans.). Cambridge, MA: MIT Press, 1986.

Cassirer, E. 1930. "Form und Technik." In *Symbol, Technik, Sprache*, edited by E. Cassirer, 39–92. Hamburg: Meiner, 1985.

Dessauer, F. 1927. *Philosophie der Technik: Das Problem der Realisierung* [Philosophy of Technology: The problem of realization]. Bonn: F. Cohen.

Drexler, E. 2003a. "Drexler counters." *Chemical & Engineering News* 81: 40–41.

Drexler, E. 2003b. "Open Letter to Richard Smalley." *Chemical & Engineering News* 81: 38–39.

Dupuy, J.-P. 2007. "Some Pitfalls in the Philosophical Foundations of nanoethics." *Journal of Medicine and Philosophy* 32 (3): 237–261.

Foucault, M. 1966. *Les Mots Et Les Choses: Une archéologie Des Sciences humaines*. Paris: Gallimard.

Geuss, R. 2010. "Realismus, Wunschdenken, Utopie." *Deutsche Zeitschrift für Philosophie* 58 (3): 419–429.

Grin, J., and A. Grunwald, eds. 2000. *Vision Assessment: Shaping Technology in 21st Century Society: Towards a Repertoire for Technology Assessment*. Heidelberg: Springer.

Jasanoff, S. 2015. "Future Imperfect: Science, Technology, and the Imaginations of Modernity." In *Dreamscapes of Modernity: Sociotechnical Imaginaries and the Fabrication of Power*, edited by S. Jasanoff and S.-H. Kim, 1–47. Chicago, IL: University of Chicago Press.

Jones, R. 2004. *Soft Machines: Nanotechnology and Life*. Oxford, England: Oxford University Press.

Jones, R. 2011. "What Has Nanotechnology Taught Us about Contemporary Technoscience?" In *Quantum Engagements: Social Reflections of Nanoscience and Emerging Technologies*, edited by T. Zülsdorf, C. Coenen, A. Ferrari, U. Fiedeler, C. Milburn and M. Wienroth, 13–26. Heidelberg: AKA.

Kearnes, M., P. Macnaghten, and J. Wilsdon. 2006. *Governing at the Nanoscale: People, Policies and Emerging Technologies*. London: Demos.

Lösch, A. 2007. "Nanoroboter und Mini-U-Boote: Mediale Vermittlung nanomedizinischer Visionen." In *Zukunftspotentiale der Nanotechnologien: Erwartungen, Anwendungen, Auswirkungen*, edited by D. Korczak and A. Lerf, 127–145. Kröning: Asanger.

Lösch, A., S. Gammel, and A. Nordmann. 2009. *Jenseits von Regulierung: Zum politischen Umgang Mit der Nanotechnologie*. Heidelberg: AKA.

Macnaghten, P., and M. Kearnes. 2007. *Deepen Deliverable 4: Working Paper: Scenario Planning and Draft Design of Focus Groups*. Unpublished. Durham, England.

Mayr, O. 1986. *Authority, Liberty and Automatic Machinery in Early Modern Europe*. Baltimore, MD: Johns Hopkins University Press.

McCray, P. 2013. *The Visioneers: How a Group of Elite Scientists Pursued Space Colonies, Nanotechnologies, and a Limitless Future*. Princeton: Princeton University Press.

Mumford, L. 1967. *The Myth of the Machine: Technics and Human Development* (*Volume 1*). New York: Harcourt Brace Jovanovich.

Mumford, L. 1970. *The Myth of the Machine: The Pentagon of Power* (*Volume 2*). London: Secker and Warburg.

Myers, N. 2008. "Conjuring machinic Life." In *Spontaneous Generations* 2 (1): 112–121.

Nerlich, B. 2005. "From Nautilus to Nanobo(a)ts: The visual Construction of Nanoscience." *AZojomo: Journal of Materials Online*. 10.2240/azojono0109.

Nordmann, A. 2008. *Technikphilosophie zur Einführung*. Hamburg: Junius.

Nordmann, A. 2010a. "A Forensics of Wishing: Technology Assessment in the Age of Technoscience." *Poeisis Prax* 7: 5–15.

Nordmann, A. 2010b. "Enhancing Material Nature." In *Nano Meets Macro: Social Perspectives on Nanoscale Sciences and Technologies*, edited by K. L. Kjølberg and F. Wickson, 283–306. Singapore: Pan Stanford.

Nordmann, A. 2013. "Visioneering Assessment: On the Construction of Tunnel Visions for Technovisionary Research and Policy." *Science, Technology and Innovation Studies* 9 (2): 89–94.

Nordmann, A. 2014. "Sanfte Technik: Vom Mythos der Maschine Zum Mythos nicht-maschineller Maschinen." In *Zur Philosophie informeller Technisierung*, edited by A. Gelhard and A. Kaminski, 21–40. Darmstadt: Wissenschaftliche Buchgesellschaft.

Nordmann, A. 2016. "Enhancing Machine Nature: From the Myth of the Machine to Sociotechnical Imaginaries of Nonmechanical Machines." In *Perfecting Human*

Futures: Transhuman Visions and Technological Imaginations, edited by B. Hurlbut and H. Tirosh-Samuelson , 195–214. Wiesbaden: Springer VS.

Nordmann, A., and A. Schwarz. 2010. "Lure of the 'Yes': The Seductive Power of Technoscience." In *Governing Future Technologies: Nanotechnology and the Rise of an Assessment Regime*, edited by M. Kaiser, M. Kurath, S. Maasen and C. Rehmann-Sutter, 255–277 (Sociology of the Sciences Yearbook 27). Dordrecht: Springer.

Nowotny, H., P. Scott, and M. Gibbons. 2004. *Wissenschaft neu denken: Wissenschaft und Öffentlichkeit in einem Zeitalter der Ungewißheit*. Weilerswist: Velbrück.

Pitt, J. 2011. *Doing Philosophy of Technology: Essays in a Pragmatist Spirit*. Dordrecht: Springer.

Rabinbach, A. 1992. *The Human Motor: Energy, Fatigue, and the Origins of Modernity*. Berkeley: University of California Press.

Reuleaux, F. 1875. "*Theoretische Kinematik. Grundzüge einer Theorie Des Maschinenwesens*. Braunschweig: Vieweg." In English: *Kinematics of Machinery: Outlines of a Theory of Machines* (A. W. B. Kennedy, Trans. & Ed.). London: Macmillan and Co., 1876.

Rip, A. 2009. "Technology as Prospective Ontology." *Synthese*, 168: 405–422.

Riskin, J. 2003. "Eighteenth-Century Wetware." *Representations* 83: 97–125.

Schwarz, A. 2009. "Green Dreams of Reason: Green Nanotechnology between Visions of Excess and Control." *Nanoethics* 3: 109–118.

Simondon, G. 1958. *On the Mode of Existence of Technical Objects*. London: University of Western Ontario.

Smalley, R. 2003a. "Smalley Responds." *Chemical & Engineering News* 81: 39–40.

Smalley, R. 2003b. "Smalley Concludes." *Chemical & Engineering News* 81: 41–42.

Taylor, C. 2004. *Modern Social Imaginaries*. Durham, NC: Duke University Press.

Index

Note: Page numbers followed by "n" refer to notes

www.ingramcontent.com/pod-product-compliance
Lightning Source LLC
LaVergne TN
LVHW021247011026
849160LV00010BA/876
* 9 7 8 1 0 3 2 3 4 4 6 6 9 *